The Principle of Least Action

Foundations and Advanced Applications

A Comprehensive Treatment from Classical Mechanics
to Quantum Field Theory and Beyond

Chocry Bendhaw, Neyl Bendhaw, Basma Bouabid

The Principle of Least Action

ISBN: 978-2-488616-00-3

Contents

Preface

The principle of least action stands as one of the most profound and unifying concepts in physics. From its humble beginnings in the study of light propagation to its central role in modern quantum field theory and string theory, this principle has demonstrated a remarkable ability to reveal the deepest structures of physical law.

This book is intended for advanced undergraduate and graduate students, as well as researchers seeking a comprehensive understanding of the principle of least action and its applications across all domains of physics. The treatment assumes familiarity with classical mechanics at the level of Goldstein, multivariable calculus, and basic linear algebra. More advanced mathematical topics are introduced as needed, with careful attention to both rigor and physical intuition.

The book's structure reflects both the historical development and the logical organization of the subject. Part I establishes the mathematical and historical foundations, tracing the evolution from Fermat's optical principle to the sophisticated formulations of Lagrange and Hamilton. Parts II and III explore applications in classical and modern physics respectively, while Part IV ventures into interdisciplinary frontiers. Part V addresses numerical methods and practical applications, recognizing that modern physics increasingly relies on computational approaches.

Throughout, we emphasize the conceptual unity that the principle of least action brings to seemingly disparate phenomena. Whether describing the path of a light ray, the orbit of a planet, or the behavior of quantum fields, nature appears to follow a profound economy principle that continues to guide our understanding of the universe.

Introduction

History and Context of the Principle of Least Action

The principle of least action represents one of physics' most elegant and powerful organizing principles. Its history spans over three centuries, beginning with Pierre de Fermat's observation about light paths in the 1650s and culminating in its modern role as the foundation for all fundamental physical theories.

The principle asserts that nature operates with a kind of economy: among all possible paths or configurations a system might take, it chooses the one that extremizes (usually minimizes) a quantity called the action. This seemingly teleological statement—that nature somehow "knows" which path to choose—has sparked philosophical debates since its inception, yet its mathematical formulation and predictive power are unassailable.

The action itself is typically defined as the time integral of the Lagrangian, a function that encodes the dynamics of the system. For a particle, the Lagrangian is usually the difference between kinetic and potential energy. The profound insight is that by requiring the action to be stationary under small variations of the path, we can derive all the equations of motion for the system.

This variational approach offers several advantages over Newton's differential formulation of mechanics. It treats constraints naturally, reveals conservation laws through symmetries (via Noether's theorem), and generalizes seamlessly to fields and relativistic systems. Perhaps most importantly, it provides a unified framework that extends from classical mechanics through quantum mechanics to quantum field theory and general relativity.

From Fermat to Hamilton, through Maupertuis, Euler and Lagrange

The historical development of the principle of least action reads like a detective story, with each generation of physicists adding crucial pieces to the puzzle. Pierre de Fermat (1601-1665) initiated this journey with his principle of least time for light propagation, stating that light travels between two points along the path requiring the least time. This principle successfully explained reflection and Snell's law of refraction, suggesting a deep connection between nature's behavior and optimization principles.

Pierre Louis Maupertuis (1698-1759) extended these ideas to mechanics, proposing that mechanical systems follow paths that minimize a quantity he called "action," defined as the product of mass, velocity, and distance. Though his formulation was imprecise and his philosophical arguments often mystical, Maupertuis correctly intuited that a universal principle governed both light and matter.

Leonhard Euler (1707-1783) provided the mathematical rigor that Maupertuis lacked. In 1744, Euler showed that the trajectories of particles could be found by solving what we now call the Euler-Lagrange equations, derived from a variational principle. His work established the calculus of variations as a mathematical discipline and demonstrated how to extract differential equations from integral principles.

Joseph-Louis Lagrange (1736-1813) revolutionized mechanics by introducing generalized coordinates and what we now call the Lagrangian formulation. His 1788 masterwork "Mécanique Analytique" presented mechanics without a single diagram, relying entirely on analytical methods. Lagrange showed that all of mechanics could be derived from a single scalar function (the Lagrangian) and a variational principle, unifying previously disparate results.

William Rowan Hamilton (1805-1865) completed this classical development by reformulating mechanics in terms of a different function (the Hamiltonian) and introducing the phase space description. His approach, published in 1834-1835, proved particularly suited to studying periodic systems and later became the natural language for quantum mechanics. Hamilton also clarified the connection between optics and mechanics, showing that both could be understood through similar variational principles.

Conceptual and Philosophical Importance of the Principle

The principle of least action raises profound philosophical questions about the nature of physical law. Its formulation appears teleological—as if particles "know" their destinations and choose paths accordingly. This troubled many physicists, including Planck, who initially viewed it as merely a mathematical convenience rather than a fundamental truth.

However, the principle's universality suggests something deeper. Every known fundamental physical theory can be formulated using an action principle: classical mechanics, electromagnetism, general relativity, quantum mechanics, and quantum field theory all emerge from appropriate action functionals. This universality hints that the principle reflects something essential about the structure of physical law itself.

From a modern perspective, we understand that the apparent teleology is illusory. The variational formulation is mathematically equivalent to local differential equations—particles don't need to "know" their destinations any more than water needs to "know" it should flow downhill. Yet the action formulation often provides insights that are obscure in the differential formulation, particularly regarding symmetries and conservation laws.

The principle also connects to deep ideas in mathematics and information theory. The path integral formulation of quantum mechanics shows that particles actually "explore" all possible paths, with the classical path emerging as the stationary phase approximation. In this view, the principle of least action describes not what nature "chooses" but what survives the sum over possibilities.

Furthermore, the action principle provides a natural framework for formulating new theories. When extending physics to new domains—whether higher energies, smaller distances, or curved spacetimes—physicists often begin by proposing an action. The requirement that physics be describable by an action principle serves as a powerful constraint on possible theories.

Part I

Mathematical and Historical Foundations

The mathematical and historical foundations of the principle of least action form the bedrock upon which all subsequent developments rest. In this part, we trace the evolution of the principle from its origins in optics through its mathematical formalization in the calculus of variations to its culmination in the Lagrangian and Hamiltonian formulations of mechanics.

Our journey begins in Chapter 1 with the historical development of variational principles, starting with Fermat's optical principle and progressing through the contributions of Maupertuis, Euler, Lagrange, and Hamilton. This historical perspective illuminates not just the mathematical development but also the conceptual struggles and insights that shaped our modern understanding.

Chapter 2 introduces the mathematical machinery of the calculus of variations, the essential tool for working with action principles. We develop the Euler-Lagrange equations systematically, explore the role of boundary conditions, and introduce Lagrange multipliers for handling constraints.

Chapters 3 and 4 present the Lagrangian and Hamiltonian formulations of mechanics respectively. These complementary approaches each offer unique insights: the Lagrangian formulation excels at handling constraints and revealing symmetries, while the Hamiltonian formulation provides the natural framework for phase space analysis and the transition to quantum mechanics.

Throughout Part I, we emphasize both mathematical rigor and physical intuition, preparing readers for the wide-ranging applications that follow in subsequent parts.

Chapter 1

Historical Formulations

The history of the principle of least action is a testament to the power of unifying ideas in physics. What began as an observation about light rays eventually encompassed all of classical mechanics and beyond. This chapter traces this remarkable development, examining the key contributions and conceptual advances that transformed a curious property of light paths into one of physics' most fundamental principles.

1.1 Fermat's Principle in Optics

Pierre de Fermat's principle of least time, proposed in 1662, marks the beginning of variational principles in physics. Fermat stated that light traveling between two points follows the path that requires the least time. Mathematically, if light travels from point A to point B through a medium with varying refractive index $n(\mathbf{r})$, it follows the path that minimizes:

$$T = \int_A^B \frac{n(\mathbf{r})}{c} \, ds \tag{1.1}$$

where ds is an element of path length and c is the speed of light in vacuum.

This principle elegantly explains the law of reflection (angle of incidence equals angle of reflection) and Snell's law of refraction ($n_1 \sin\theta_1 = n_2 \sin\theta_2$). For reflection, all paths except the symmetric one require longer times. For refraction, the path "breaks" at the interface to balance the time spent in each medium optimally.

Fermat's insight was revolutionary for several reasons. First, it provided a unified explanation for seemingly different phenomena. Second, it introduced an integral principle—the path is determined by a global property (total time) rather than local conditions. Third, it suggested that nature operates according to an economy principle, choosing optimal paths among all possibilities.

However, Fermat's principle also raised philosophical concerns. How does light "know" which path to take? The principle seems to require that light explores all possible paths before choosing the quickest. This teleological aspect troubled many natural philosophers and would remain controversial until the development of quantum mechanics provided a startling resolution: in Feynman's path integral formulation, light actually does explore all paths, with the classical path emerging from constructive interference.

The mathematical techniques for solving problems using Fermat's principle were initially limited. Fermat himself used geometric arguments for simple cases. The full power of the variational approach would not be realized until Euler developed the calculus of variations in the following century.

Nevertheless, Fermat's principle established the template for all subsequent action principles: identify a quantity that nature extremizes, then derive the observed behavior from this extremization condition.

1.2 The Works of Maupertuis and the Concept of Action

Pierre Louis Maupertuis extended variational principles from optics to mechanics in the 1740s, introducing the concept of "action" that gives the principle its modern name. Maupertuis was motivated by both scientific and philosophical considerations, believing he had discovered a universal principle that revealed God's economy in nature.

Maupertuis defined action for a particle as the product of mass, velocity, and distance traveled:

$$\text{Action} = \int mv\, ds \tag{1.2}$$

He proposed that among all possible paths between two points, a particle follows the path that minimizes this action. For a free particle, this yields a straight line—the shortest path. For more complex situations, Maupertuis' formulation was often vague and sometimes incorrect in detail.

Despite its limitations, Maupertuis' principle represented a crucial conceptual advance. He recognized that mechanical systems, like optical ones, follow variational principles. He also introduced the term "action," though his definition differed from the modern one. Most importantly, he promoted the idea that a single principle might govern all physical phenomena.

Maupertuis' work sparked fierce controversies. Samuel König claimed that Leibniz had priority, leading to a bitter dispute that divided the European scientific community. The Berlin Academy, under Maupertuis' presidency, sided with him, but the controversy damaged his reputation. Additionally, Maupertuis' theological arguments for the principle—that it demonstrated God's wisdom in creating the most economical universe—mixed science with metaphysics in ways that made many uncomfortable.

The controversy obscured the genuine insights in Maupertuis' work. His principle, while imperfectly formulated, correctly captured the variational nature of mechanics. His vision of a unified principle governing diverse phenomena proved prescient. And his introduction of "action" as the key quantity, even if imprecisely defined, pointed toward the modern formulation.

Maupertuis also made specific applications of his principle. He used it to derive the law of refraction, recovering Fermat's result from a mechanical perspective. He applied it to collision problems, though with mixed success due to his imprecise formulation. These applications, despite their flaws, demonstrated the potential universality of variational principles.

1.3 Contributions of Euler and Lagrange

Leonhard Euler transformed Maupertuis' philosophical vision into precise mathematics. In 1744, Euler published "Methodus inveniendi lineas curvas maximi minimive proprietate gaudentes" (Method of finding curved lines enjoying maximum or minimum properties), founding the calculus of variations and providing the mathematical tools needed to implement variational principles rigorously.

Euler's key insight was to consider how integrals change under small variations of the path. For an integral of the form:

$$I = \int F(x, y, y')\, dx \tag{1.3}$$

Euler derived the condition for extremization, now known as the Euler-Lagrange equation:

$$\frac{d}{dx}\left(\frac{\partial F}{\partial y'}\right) - \frac{\partial F}{\partial y} = 0 \tag{1.4}$$

This differential equation determines the extremal path $y(x)$. Euler showed how to apply this to mechanical problems, deriving Newton's equations from variational principles.

Euler also clarified Maupertuis' principle, showing that for conservative systems, the quantity to be extremized should be:

$$\int 2T\, dt = \int mv^2\, dt \tag{1.5}$$

where T is kinetic energy. This formulation correctly yields Newton's laws and avoids the ambiguities in Maupertuis' original statement.

Joseph-Louis Lagrange revolutionized mechanics by systematically developing what we now call the Lagrangian formulation. His 1788 treatise "Mécanique Analytique" presented mechanics as a branch of analysis, deriving everything from variational principles without geometric reasoning.

Lagrange's crucial innovations included:

1. **Generalized coordinates**: Lagrange showed that mechanical systems could be described using any convenient set of coordinates $q_1, q_2, \ldots, q_n$, not just Cartesian coordinates. This dramatically simplified problems with constraints.

2. **The Lagrangian**: Lagrange identified the key function $L = T - V$ (kinetic minus potential energy) and showed that the equations of motion follow from extremizing the action:

$$S = \int L(q, \dot{q}, t)\, dt \tag{1.6}$$

3. **Systematic treatment of constraints**: Lagrange developed methods for handling both holonomic constraints (restrictions on coordinates) and non-holonomic constraints (restrictions on velocities) using multiplier techniques.

4. **Virtual work principle**: Lagrange connected variational principles to the principle of virtual work, providing an alternative derivation of his equations.

The Lagrangian formulation offered tremendous advantages. Complex problems involving constraints, which were difficult in Newton's formulation, became straightforward. The formulation was manifestly coordinate-independent. Most importantly, it revealed the deep structure of mechanics, showing how dynamics emerges from a single scalar function.

Lagrange's approach also clarified the relationship between symmetries and conservation laws, though the full understanding would await Emmy Noether's theorem in the 20th century. Lagrange showed that if the Lagrangian doesn't depend on a particular coordinate (a cyclic coordinate), the corresponding momentum is conserved.

1.4 Hamiltonian Formulation

William Rowan Hamilton completed the classical development of analytical mechanics in the 1830s by reformulating it in terms of a different function and introducing phase space methods. Hamilton's approach, while mathematically equivalent to Lagrange's, offered new insights and proved particularly suitable for certain classes of problems.

Hamilton's starting point was a transformation of variables. Instead of working with coordinates q and velocities $\dot{q}$, Hamilton used coordinates q and momenta $p = \partial L/\partial \dot{q}$. This required transforming from the Lagrangian $L(q, \dot{q}, t)$ to a new function, the Hamiltonian:

$$H(q, p, t) = \sum_i p_i \dot{q}_i - L \tag{1.7}$$

where $\dot{q}_i$ is expressed in terms of q and p by inverting the momentum definition.

Hamilton showed that the equations of motion take a remarkably symmetric form:

$$\dot{q}_i = \frac{\partial H}{\partial p_i} \tag{1.8}$$

$$\dot{p}_i = -\frac{\partial H}{\partial q_i} \tag{1.9}$$

These are Hamilton's canonical equations. The symmetry between q and p revealed a deep structure in mechanics that would prove crucial for later developments.

Hamilton's formulation offered several advantages:

1. **Phase space perspective**: The $2n$-dimensional space of coordinates and momenta (phase space) provides a geometric view of dynamics. Trajectories are curves in phase space, and the flow preserves phase space volume (Liouville's theorem).

2. **Canonical transformations**: Hamilton developed a theory of transformations that preserve the form of his equations. These canonical transformations include both point transformations and more general changes of variables, providing powerful solution techniques.

3. **Connection to optics**: Hamilton discovered a profound analogy between mechanics and optics. Just as light rays are perpendicular to wavefronts, particle trajectories are perpendicular to surfaces of constant action. This mechanical-optical analogy would later inspire Schrödinger in developing wave mechanics.

4. **Hamilton-Jacobi theory**: Hamilton and Carl Gustav Jacob Jacobi developed a method for solving mechanical problems by finding a generating function that transforms to coordinates where all momenta are constants. This approach connects classical and quantum mechanics.

5. **Natural framework for perturbation theory**: The Hamiltonian formulation proves particularly suitable for studying small deviations from integrable systems, a crucial technique in celestial mechanics and later in quantum mechanics.

Hamilton's work also introduced important mathematical structures. The Poisson bracket:

$$\{f, g\} = \sum_i \left(\frac{\partial f}{\partial q_i} \frac{\partial g}{\partial p_i} - \frac{\partial f}{\partial p_i} \frac{\partial g}{\partial q_i} \right) \tag{1.10}$$

provides an algebraic structure on phase space functions that directly translates to the commutator in quantum mechanics.

The Hamiltonian formulation completed the classical development of analytical mechanics. From Fermat's simple observation about light paths, through Maupertuis' vision of universal economy, Euler's mathematical rigor, and Lagrange's systematic reformulation, to Hamilton's phase space methods, the principle of least action had evolved into a complete framework for classical physics.

This framework would prove robust enough to extend into relativistic and quantum domains, testifying to the profound insight captured by the variational approach to physics.

Chapter 2

Calculus of Variations

The calculus of variations provides the mathematical foundation for the principle of least action. Just as differential calculus deals with finding extrema of functions, variational calculus addresses the extremization of functionals—mappings from functions to numbers. This chapter develops the essential tools systematically, beginning with the fundamental concepts and building to sophisticated techniques for handling constraints.

The importance of variational methods extends far beyond mechanics. They appear in optics (Fermat's principle), electromagnetism (electromagnetic action), quantum field theory (path integrals), general relativity (Einstein-Hilbert action), and even economics and biology. Mastering these techniques opens doors to understanding nature's variational principles across all scales.

2.1 Introduction to Variational Calculus

2.1.1 Functionals and Their Variations

A functional is a mapping that assigns a number to each function in some class. In physics, the action is a functional: given a path $q(t)$, it returns a number $S[q]$. The central problem of variational calculus is finding functions that extremize functionals.

Consider a functional of the form:

$$J[y] = \int_{x_1}^{x_2} F(x, y(x), y'(x))\, dx \tag{2.1}$$

where F is a given function of three variables, $y(x)$ is the unknown function we seek, and $y'(x) = dy/dx$. The functional J assigns to each function $y(x)$ a number obtained by integrating F along the curve $y(x)$.

Example 2.1. The arc length of a curve $y(x)$ from x_1 to x_2 is:

$$L[y] = \int_{x_1}^{x_2} \sqrt{1 + (y')^2}\, dx \tag{2.2}$$

Here $F(x, y, y') = \sqrt{1 + (y')^2}$, which doesn't depend explicitly on x or y.

To find extrema of functionals, we need the concept of variation. Consider a family of functions:

$$y(x, \varepsilon) = y(x) + \varepsilon\eta(x) \tag{2.3}$$

where $y(x)$ is our candidate extremal function, $\eta(x)$ is an arbitrary smooth function satisfying $\eta(x_1) = \eta(x_2) = 0$ (maintaining fixed endpoints), and ε is a small parameter. The variation of y is defined as:

$$\delta y = \left.\frac{\partial y}{\partial \varepsilon}\right|_{\varepsilon=0} = \eta(x) \tag{2.4}$$

This represents an infinitesimal change in the function $y(x)$.

2.1.2 The Fundamental Lemma of Variational Calculus

A crucial tool in deriving the Euler-Lagrange equation is the fundamental lemma:

Lemma 2.1 (Fundamental Lemma). If $\int_{x_1}^{x_2} g(x)\eta(x)dx = 0$ for all smooth functions $\eta(x)$ vanishing at x_1 and x_2, then $g(x) = 0$ for $x_1 < x < x_2$.

Proof. Suppose $g(x_0) \neq 0$ at some point $x_0 \in (x_1, x_2)$. By continuity, $g(x) \neq 0$ in some interval around x_0. Choose $\eta(x)$ to be positive in this interval and zero elsewhere (smoothly connected). Then $\int g(x)\eta(x)dx > 0$ if $g(x_0) > 0$, contradicting our hypothesis. Therefore, $g(x) = 0$ everywhere. □

This lemma is fundamental because it allows us to conclude that the integrand itself vanishes when we know only that certain integrals vanish.

2.1.3 First Variation and Extremization Condition

For a functional $J[y]$ to have an extremum at $y(x)$, we require:

$$\delta J = \left.\frac{dJ[y+\varepsilon\eta]}{d\varepsilon}\right|_{\varepsilon=0} = 0 \tag{2.5}$$

for all admissible variations $\eta(x)$. This is the variational analogue of requiring $df/dx = 0$ for extrema of ordinary functions.

Computing the first variation:

$$\delta J = \left.\frac{d}{d\varepsilon}\int_{x_1}^{x_2} F(x, y+\varepsilon\eta, y'+\varepsilon\eta')\,dx\right|_{\varepsilon=0} \tag{2.6}$$

$$= \int_{x_1}^{x_2}\left[\frac{\partial F}{\partial y}\eta + \frac{\partial F}{\partial y'}\eta'\right]dx \tag{2.7}$$

where we've used the chain rule and interchanged differentiation and integration (justified under suitable regularity conditions).

2.2 Euler-Lagrange Equation

2.2.1 Derivation of the Euler-Lagrange Equation

To proceed, we integrate the second term by parts:

$$\int_{x_1}^{x_2}\frac{\partial F}{\partial y'}\eta'\,dx = \left[\frac{\partial F}{\partial y'}\eta\right]_{x_1}^{x_2} - \int_{x_1}^{x_2}\frac{d}{dx}\left(\frac{\partial F}{\partial y'}\right)\eta\,dx \tag{2.8}$$

Since $\eta(x_1) = \eta(x_2) = 0$, the boundary term vanishes. Therefore:

$$\delta J = \int_{x_1}^{x_2}\left[\frac{\partial F}{\partial y} - \frac{d}{dx}\left(\frac{\partial F}{\partial y'}\right)\right]\eta(x)\,dx \tag{2.9}$$

For $\delta J = 0$ for all $\eta(x)$, the fundamental lemma requires:

$$\frac{\partial F}{\partial y} - \frac{d}{dx}\left(\frac{\partial F}{\partial y'}\right) = 0 \tag{2.10}$$

This is the celebrated Euler-Lagrange equation. Any function $y(x)$ that extremizes $J[y]$ must satisfy this differential equation.

2.2.2 Structure and Interpretation

The Euler-Lagrange equation is generally a second-order differential equation. Expanding the total derivative:

$$\frac{\partial F}{\partial y} - \frac{\partial^2 F}{\partial x \partial y'} - \frac{\partial^2 F}{\partial y \partial y'} y' - \frac{\partial^2 F}{\partial y'^2} y'' = 0 \tag{2.11}$$

The highest derivative y'' appears linearly if $\partial^2 F/\partial y'^2 \neq 0$, which is the regular case. The equation then has the form:

$$y'' = G(x, y, y') \tag{2.12}$$

guaranteeing local existence and uniqueness of solutions given initial conditions $y(x_0)$ and $y'(x_0)$.

Physical Interpretation: In mechanics, F is the Lagrangian $L = T - V$. The term $\partial L/\partial q$ represents generalized forces, while $d/dt(\partial L/\partial \dot{q})$ represents the rate of change of generalized momentum. The Euler-Lagrange equation thus generalizes Newton's $F = ma$.

2.2.3 Examples and Applications

Example 2.2 (Shortest Path (Geodesic))**.** For $F = \sqrt{1 + (y')^2}$, we have:

$$\frac{\partial F}{\partial y} = 0 \tag{2.13}$$

$$\frac{\partial F}{\partial y'} = \frac{y'}{\sqrt{1 + (y')^2}} \tag{2.14}$$

$$\frac{d}{dx}\left(\frac{\partial F}{\partial y'}\right) = \frac{d}{dx}\left[\frac{y'}{\sqrt{1 + (y')^2}}\right] = 0 \tag{2.15}$$

This yields $y'' = 0$, so $y = ax + b$, confirming that straight lines are the shortest paths in Euclidean space.

Example 2.3 (Brachistochrone Problem)**.** Find the curve of fastest descent under gravity. The time functional is:

$$T[y] = \frac{1}{\sqrt{2g}} \int_{x_1}^{x_2} \sqrt{\frac{1 + (y')^2}{y}}\, dx \tag{2.16}$$

The Euler-Lagrange equation yields a cycloid—the curve traced by a point on a rolling circle.

Example 2.4 (Minimal Surface). For a surface of revolution, the area functional is:

$$A[y] = 2\pi \int_{x_1}^{x_2} y\sqrt{1 + (y')^2}\, dx \tag{2.17}$$

The Euler-Lagrange equation gives:

$$y\sqrt{1 + (y')^2} - \frac{y(y')^2}{\sqrt{1 + (y')^2}} = c \tag{2.18}$$

where c is a constant. The solution is a catenary: $y = c\cosh((x - x_0)/c)$.

2.2.4 Multiple Dependent Variables

When the functional depends on several functions:

$$J[y_1, y_2, \ldots, y_n] = \int_{x_1}^{x_2} F(x, y_1, \ldots, y_n, y_1', \ldots, y_n')\, dx \tag{2.19}$$

we obtain a system of Euler-Lagrange equations:

$$\frac{\partial F}{\partial y_i} - \frac{d}{dx}\left(\frac{\partial F}{\partial y_i'}\right) = 0, \quad i = 1, 2, \ldots, n \tag{2.20}$$

Each function $y_i(x)$ must satisfy its corresponding equation.

2.2.5 Higher Derivatives

For functionals involving higher derivatives:

$$J[y] = \int_{x_1}^{x_2} F(x, y, y', y'', \ldots, y^{(n)})\, dx \tag{2.21}$$

the Euler-Lagrange equation generalizes to:

$$\frac{\partial F}{\partial y} - \frac{d}{dx}\left(\frac{\partial F}{\partial y'}\right) + \frac{d^2}{dx^2}\left(\frac{\partial F}{\partial y''}\right) - \ldots + (-1)^n \frac{d^n}{dx^n}\left(\frac{\partial F}{\partial y^{(n)}}\right) = 0 \tag{2.22}$$

Each integration by parts introduces a sign change, leading to the alternating pattern.

2.3 Lagrange Multipliers

2.3.1 Constrained Variational Problems

Physical systems often involve constraints—restrictions on the allowed functions. For example, a pendulum's bob must maintain constant distance from the pivot, or a bead must slide along a wire. The method of Lagrange multipliers extends naturally from ordinary calculus to variational problems.

Consider extremizing $J[y]$ subject to the constraint:

$$K[y] = \int_{x_1}^{x_2} G(x, y, y')\, dx = c \tag{2.23}$$

where c is a constant. We form the augmented functional:

$$\tilde{J}[y, \lambda] = J[y] - \lambda(K[y] - c) \tag{2.24}$$

where λ is the Lagrange multiplier (a constant, not a function).

2.3.2 Isoperimetric Problems

The classic isoperimetric problem asks: among all curves of fixed length, which encloses maximum area? Mathematically, maximize:

$$A[y] = \frac{1}{2}\int_{x_1}^{x_2} y\, dx \tag{2.25}$$

subject to:

$$L[y] = \int_{x_1}^{x_2} \sqrt{1 + (y')^2}\, dx = l \tag{2.26}$$

The augmented functional is:

$$\tilde{J}[y] = \frac{1}{2}\int_{x_1}^{x_2} y\, dx - \lambda \int_{x_1}^{x_2} \sqrt{1 + (y')^2}\, dx \tag{2.27}$$

Leading to the Euler-Lagrange equation:

$$\frac{1}{2} - \lambda \frac{d}{dx}\left[\frac{y'}{\sqrt{1 + (y')^2}}\right] = 0 \tag{2.28}$$

The solution is a circular arc, confirming the ancient result that circles maximize area for given perimeter.

2.3.3 Multiple Constraints

For multiple integral constraints:

$$K_j[y] = \int_{x_1}^{x_2} G_j(x, y, y')\, dx = c_j, \quad j = 1, 2, \ldots, m \tag{2.29}$$

we introduce m multipliers:

$$\tilde{J}[y, \lambda_1, \ldots, \lambda_m] = J[y] - \sum_j \lambda_j (K_j[y] - c_j) \tag{2.30}$$

The extremal function must satisfy:

$$\frac{\partial}{\partial y}\left(F - \sum_j \lambda_j G_j\right) - \frac{d}{dx}\frac{\partial}{\partial y'}\left(F - \sum_j \lambda_j G_j\right) = 0 \tag{2.31}$$

along with the m constraint equations.

2.3.4 Differential Constraints

Some constraints involve the functions pointwise rather than integrally:

$$g(x, y_1, \ldots, y_n, y_1', \ldots, y_n') = 0 \tag{2.32}$$

For such holonomic constraints, we use a Lagrange multiplier function $\lambda(x)$:

$$\tilde{J}[y_1, \ldots, y_n, \lambda] = \int_{x_1}^{x_2} [F(x, y_1, \ldots, y_n, y_1', \ldots, y_n') - \lambda(x) g(x, y_1, \ldots, y_n, y_1', \ldots, y_n')]\, d\text{:} \tag{2.33}$$

This yields $n + 1$ equations: n Euler-Lagrange equations for the y_i and the constraint equation for $\lambda(x)$.

2.3.5 Example: Geodesics on Surfaces

To find shortest paths on a surface defined by $g(x, y, z) = 0$, we minimize arc length:

$$L = \int \sqrt{\dot{x}^2 + \dot{y}^2 + \dot{z}^2}\, dt \tag{2.34}$$

subject to $g(x(t), y(t), z(t)) = 0$. Using a multiplier function $\lambda(t)$:

$$\tilde{L} = \int [\sqrt{\dot{x}^2 + \dot{y}^2 + \dot{z}^2} - \lambda(t) g(x, y, z)]\, dt \tag{2.35}$$

The resulting equations are:

$$\frac{d}{dt}\left(\frac{\dot{x}}{\sqrt{\dot{x}^2 + \dot{y}^2 + \dot{z}^2}}\right) = \lambda \frac{\partial g}{\partial x} \tag{2.36}$$

with similar equations for y and z, plus the constraint $g = 0$.

2.4 Boundary Conditions and Holonomic Constraints

2.4.1 Natural Boundary Conditions

So far, we've assumed fixed endpoints: $y(x_1) = y_1$ and $y(x_2) = y_2$. However, some problems have free endpoints or other boundary conditions. When endpoints can vary, additional conditions emerge from the variational principle.

Consider varying the endpoint x_2 while keeping $y(x_2)$ free. The variation of J includes a boundary term:

$$\delta J = \int_{x_1}^{x_2} \left[\frac{\partial F}{\partial y} - \frac{d}{dx}\left(\frac{\partial F}{\partial y'}\right)\right] \delta y\, dx + \left[F - y'\frac{\partial F}{\partial y'}\right]_{x=x_2} \delta x_2 + \left[\frac{\partial F}{\partial y'}\right]_{x=x_2} \delta y_2 \tag{2.37}$$

For a genuine extremum, all terms must vanish independently:

1. The Euler-Lagrange equation in the interior
2. Transversality condition: $\left[F - y'\frac{\partial F}{\partial y'}\right]_{x=x_2} = 0$
3. Natural boundary condition: $\left[\frac{\partial F}{\partial y'}\right]_{x=x_2} = 0$ (if $y(x_2)$ is free)

2.4.2 Mixed Boundary Conditions

Physical problems often involve mixed conditions. For example:

- Fixed position at one end, free at the other

- Fixed position but free slope
- Prescribed relationship between position and slope

Each case requires careful analysis of boundary terms. The general principle: terms arising from integration by parts must vanish either because the variation vanishes (fixed quantities) or because their coefficients vanish (natural conditions).

2.4.3 Holonomic Constraints

Holonomic constraints are restrictions on coordinates that can be expressed as:

$$f_j(x, y_1, \ldots, y_n) = 0, \quad j = 1, \ldots, m < n \tag{2.38}$$

These reduce the system's degrees of freedom from n to $n - m$. Two approaches exist:

Method 1: Elimination Solve the constraints for m variables in terms of the remaining $n - m$, then substitute into the functional. This yields an unconstrained problem in fewer variables.

Method 2: Lagrange Multipliers Retain all n variables but introduce m multiplier functions $\lambda_j(x)$:

$$\tilde{L} = L - \sum_j \lambda_j(x) f_j(x, y_1, \ldots, y_n) \tag{2.39}$$

This yields $n + m$ equations but often proves more systematic.

2.4.4 Non-Holonomic Constraints

Non-holonomic constraints involve velocities and cannot be integrated to position constraints:

$$g(x, y_1, \ldots, y_n, y_1', \ldots, y_n') = 0 \tag{2.40}$$

These arise in rolling constraints (wheels rolling without slipping) and other mechanical systems. The variational treatment requires special care since the constraint restricts the allowed variations.

2.4.5 Example: Particle on a Sphere

Consider a particle constrained to move on a sphere $x^2 + y^2 + z^2 = R^2$. The Lagrangian is:

$$L = \frac{m}{2}(\dot{x}^2 + \dot{y}^2 + \dot{z}^2) - V(x, y, z) \tag{2.41}$$

Using a Lagrange multiplier:

$$\tilde{L} = \frac{m}{2}(\dot{x}^2 + \dot{y}^2 + \dot{z}^2) - V(x, y, z) - \lambda(x^2 + y^2 + z^2 - R^2) \tag{2.42}$$

The equations of motion are:

$$m\ddot{x} = -\frac{\partial V}{\partial x} - 2\lambda x \tag{2.43}$$

with similar equations for y and z. The multiplier λ represents the normal force keeping the particle on the sphere.

2.4.6 Vakonomic vs. Nonholonomic Dynamics

A subtle issue arises with velocity-dependent constraints. The vakonomic approach treats them like holonomic constraints using multipliers, while the nonholonomic approach recognizes that they don't reduce the configuration space. The two methods generally yield different equations of motion.

Physical systems with rolling constraints are genuinely nonholonomic and require the latter approach.

2.4.7 Conservation Laws and Constraints

Constraints often lead to modified conservation laws. For a system with symmetry, Noether's theorem yields conserved quantities. Constraints can:

- Break symmetries, destroying conservation laws
- Introduce new symmetries specific to the constrained system
- Modify the form of conserved quantities

Understanding this interplay is crucial for analyzing constrained systems.

Summary

The calculus of variations provides the mathematical foundation for physics' variational principles. Starting from the concept of functionals and their variations, we derived the Euler-Lagrange equation—the fundamental tool for finding extremal functions. The method of Lagrange multipliers extends these techniques to constrained problems, while careful analysis of boundary conditions ensures well-posed problems.

These mathematical tools, developed in the 18th and 19th centuries for classical mechanics, have proven remarkably durable. They apply unchanged to field theories, where functions of one variable become fields depending on spacetime coordinates. They extend to quantum mechanics through path integrals. They even appear in optimal control theory and machine learning.

The key insight is that nature's laws often take variational form: rather than prescribing local rules, physics identifies global quantities that must be extremized. This variational structure reveals deep symmetries and provides powerful solution methods. As we'll see in subsequent chapters, the Lagrangian and Hamiltonian formulations of mechanics exploit these variational principles to provide profound insights into the structure of physical law.

Chapter 3

Lagrangian Formalism

The Lagrangian formalism represents one of the most elegant and powerful approaches to classical mechanics. Developed by Joseph-Louis Lagrange in the late 18th century, this formulation transforms mechanics from a vector-based discipline focused on forces and accelerations to a scalar-based approach centered on energy functions and generalized coordinates. This transformation not only simplifies many mechanical problems but also reveals deep connections between symmetries and conservation laws, provides natural frameworks for constrained systems, and extends seamlessly to field theories and quantum mechanics.

The genius of Lagrange's approach lies in its coordinate independence and its ability to handle constraints naturally. While Newton's formulation requires careful accounting of all forces, including constraint forces, the Lagrangian method automatically incorporates constraints through the choice of coordinates. This chapter develops the Lagrangian formalism systematically, beginning with the concept of generalized coordinates and culminating in Noether's profound theorem linking symmetries to conservation laws.

3.1 Generalized Coordinates

3.1.1 Configuration Space and Degrees of Freedom

The foundation of Lagrangian mechanics rests on the concept of configuration space—the space of all possible positions of a mechanical system. For a system of N particles in three-dimensional space, the configuration is specified by $3N$ Cartesian coordinates. However, constraints often reduce the effective dimensionality.

Consider a rigid body: although composed of many particles, only six coordinates are needed to specify its configuration—three for the center of mass position and three for orientation. The number of independent coordinates needed to specify a system's configuration is called the number of degrees of freedom.

Definition 3.1. A system with n degrees of freedom can be described by n independent generalized coordinates $q_1, q_2, \ldots, q_n$, collectively denoted as $\mathbf{q} = (q_1, q_2, \ldots, q_n)$.

The term "generalized" emphasizes that these coordinates need not be Cartesian or even have dimensions of length. They can be angles, areas, or any parameters that uniquely specify the configuration.

3.1.2 Examples of Generalized Coordinates

Example 3.1 (Spherical Pendulum). A mass m suspended by a rigid rod of length l from a fixed point can swing in three dimensions. Although the mass moves in 3D space, the constraint $|\mathbf{r}| = l$ reduces the degrees of freedom to two. Natural generalized coordinates are:

$$q_1 = \theta \quad \text{(polar angle from vertical)} \tag{3.1}$$
$$q_2 = \phi \quad \text{(azimuthal angle)} \tag{3.2}$$

The Cartesian coordinates are expressed as:

$$x = l \sin\theta \cos\phi \tag{3.3}$$
$$y = l \sin\theta \sin\phi \tag{3.4}$$
$$z = -l \cos\theta \tag{3.5}$$

Example 3.2 (Double Pendulum). Two masses connected by rigid rods form a double pendulum. Despite motion in a plane (seemingly 4 coordinates for 2 masses), constraints reduce this to two degrees of freedom:

$$q_1 = \theta_1 \quad \text{(angle of first rod from vertical)} \tag{3.6}$$
$$q_2 = \theta_2 \quad \text{(angle of second rod from vertical)} \tag{3.7}$$

Example 3.3 (Particle on a Surface). For a particle constrained to move on a surface $f(x, y, z) = 0$, two coordinates suffice. If the surface is a sphere $x^2 + y^2 + z^2 = R^2$, natural choices are spherical angles (θ, ϕ). For more complex surfaces, any parametrization (u, v) works.

3.1.3 Constraints and Their Classification

Constraints are conditions that restrict the system's motion. They are classified as:

Holonomic Constraints: Expressible as equations involving only coordinates and time:

$$f(q_1, \ldots, q_n, t) = 0 \tag{3.8}$$

Examples include rigid body constraints, particles on surfaces, and fixed-length connections.

Non-holonomic Constraints: Involve velocities and cannot be integrated to purely positional constraints:

$$g(q_1, \ldots, q_n, \dot{q}_1, \ldots, \dot{q}_n, t) = 0 \tag{3.9}$$

The classic example is a rolling wheel: the no-slip condition $v = r\omega$ relates velocity to angular velocity but doesn't restrict the wheel's final position.

Scleronomic vs. Rheonomic:

- Scleronomic: time-independent constraints
- Rheonomic: explicitly time-dependent constraints

3.1.4 Virtual Displacements

A virtual displacement δq is an infinitesimal change in coordinates consistent with the constraints at a fixed instant of time. Unlike actual displacements dq that occur over time dt, virtual displacements are instantaneous mathematical variations.

For holonomic constraints $f(\mathbf{q}, t) = 0$, virtual displacements satisfy:

$$\sum_i \frac{\partial f}{\partial q_i} \delta q_i = 0 \tag{3.10}$$

Note that $\partial f/\partial t$ doesn't appear because time is held fixed.

The principle of virtual work states that for a system in equilibrium, the virtual work done by all forces is zero:

$$\delta W = \sum_i \mathbf{F}_i \cdot \delta \mathbf{r}_i = 0 \tag{3.11}$$

This principle, combined with D'Alembert's principle for dynamics, leads directly to Lagrange's equations.

3.1.5 Transformation from Cartesian to Generalized Coordinates

The relationship between Cartesian coordinates $\mathbf{r}_i$ and generalized coordinates $\mathbf{q}$ is:

$$\mathbf{r}_i = \mathbf{r}_i(q_1, \ldots, q_n, t) \tag{3.12}$$

Velocities transform as:

$$\dot{\mathbf{r}}_i = \sum_j \frac{\partial \mathbf{r}_i}{\partial q_j}\dot{q}_j + \frac{\partial \mathbf{r}_i}{\partial t} \tag{3.13}$$

The kinetic energy, originally $T = \frac{1}{2}\sum_i m_i \dot{\mathbf{r}}_i^2$, becomes a function of $\mathbf{q}$, $\dot{\mathbf{q}}$, and possibly t:

$$T = T(\mathbf{q}, \dot{\mathbf{q}}, t) \tag{3.14}$$

Crucially, for scleronomic constraints, T is a homogeneous quadratic function of the generalized velocities:

$$T = \frac{1}{2}\sum_{ij} M_{ij}(\mathbf{q})\dot{q}_i\dot{q}_j \tag{3.15}$$

where $M_{ij}(\mathbf{q})$ is the mass matrix, encoding the system's inertial properties in generalized coordinates.

3.2 Derivation of Lagrange Equations

3.2.1 D'Alembert's Principle and Virtual Work

D'Alembert's principle extends the principle of virtual work to dynamics by introducing inertial forces. For a particle of mass m:

$$(\mathbf{F} - m\mathbf{a}) \cdot \delta\mathbf{r} = 0 \tag{3.16}$$

Summing over all particles:

$$\sum_i (\mathbf{F}_i - m_i\ddot{\mathbf{r}}_i) \cdot \delta\mathbf{r}_i = 0 \tag{3.17}$$

This can be separated into applied forces and inertial terms:

$$\sum_i \mathbf{F}_i \cdot \delta\mathbf{r}_i - \sum_i m_i\ddot{\mathbf{r}}_i \cdot \delta\mathbf{r}_i = 0 \tag{3.18}$$

3.2.2 Transformation to Generalized Coordinates

Express the virtual displacements in terms of generalized coordinates:

$$\delta \mathbf{r}_i = \sum_j \frac{\partial \mathbf{r}_i}{\partial q_j} \delta q_j \tag{3.19}$$

The virtual work becomes:

$$\delta W = \sum_i \mathbf{F}_i \cdot \delta \mathbf{r}_i = \sum_j Q_j \delta q_j \tag{3.20}$$

where the generalized forces are:

$$Q_j = \sum_i \mathbf{F}_i \cdot \frac{\partial \mathbf{r}_i}{\partial q_j} \tag{3.21}$$

For the inertial term, we use a clever manipulation. Note that:

$$\frac{d}{dt}\left(\frac{\partial \mathbf{r}_i}{\partial q_j}\right) = \frac{\partial \dot{\mathbf{r}}_i}{\partial q_j} \tag{3.22}$$

Therefore:

$$m_i \ddot{\mathbf{r}}_i \cdot \frac{\partial \mathbf{r}_i}{\partial q_j} = \frac{d}{dt}\left[m_i \dot{\mathbf{r}}_i \cdot \frac{\partial \mathbf{r}_i}{\partial q_j}\right] - m_i \dot{\mathbf{r}}_i \cdot \frac{d}{dt}\left(\frac{\partial \mathbf{r}_i}{\partial q_j}\right) \tag{3.23}$$

$$= \frac{d}{dt}\left[m_i \dot{\mathbf{r}}_i \cdot \frac{\partial \dot{\mathbf{r}}_i}{\partial \dot{q}_j}\right] - m_i \dot{\mathbf{r}}_i \cdot \frac{\partial \dot{\mathbf{r}}_i}{\partial q_j} \tag{3.24}$$

3.2.3 The Lagrange Equations

Using $T = \frac{1}{2}\sum_i m_i \dot{\mathbf{r}}_i^2$, we find:

$$\frac{\partial T}{\partial \dot{q}_j} = \sum_i m_i \dot{\mathbf{r}}_i \cdot \frac{\partial \dot{\mathbf{r}}_i}{\partial \dot{q}_j} = \sum_i m_i \dot{\mathbf{r}}_i \cdot \frac{\partial \mathbf{r}_i}{\partial q_j} \tag{3.25}$$

$$\frac{\partial T}{\partial q_j} = \sum_i m_i \dot{\mathbf{r}}_i \cdot \frac{\partial \dot{\mathbf{r}}_i}{\partial q_j} \tag{3.26}$$

Substituting into D'Alembert's principle:

$$\sum_j \left[\frac{d}{dt}\left(\frac{\partial T}{\partial \dot{q}_j}\right) - \frac{\partial T}{\partial q_j} - Q_j\right] \delta q_j = 0 \tag{3.27}$$

Since the δq_j are independent, each coefficient must vanish:

$$\frac{d}{dt}\left(\frac{\partial T}{\partial \dot{q}_j}\right) - \frac{\partial T}{\partial q_j} = Q_j \tag{3.28}$$

These are Lagrange's equations for systems with non-conservative forces.

3.2.4 Conservative Systems and the Lagrangian

For conservative forces derivable from a potential $V(\mathbf{q}, t)$:

$$\mathbf{F}_i = -\nabla_i V \tag{3.29}$$

The generalized forces become:

$$Q_j = -\frac{\partial V}{\partial q_j} \tag{3.30}$$

Define the Lagrangian:

$$L = T - V \tag{3.31}$$

Lagrange's equations take the elegant form:

$$\frac{d}{dt}\left(\frac{\partial L}{\partial \dot{q}_j}\right) - \frac{\partial L}{\partial q_j} = 0 \tag{3.32}$$

This is the cornerstone of analytical mechanics. From a single scalar function L, all equations of motion follow.

3.2.5 Examples and Applications

Example 3.4 (Simple Pendulum). For a pendulum of length l:

$$T = \frac{1}{2}ml^2\dot{\theta}^2 \tag{3.33}$$

$$V = -mgl\cos\theta \tag{3.34}$$

$$L = \frac{1}{2}ml^2\dot{\theta}^2 + mgl\cos\theta \tag{3.35}$$

The Lagrange equation:

$$\frac{d}{dt}(ml^2\dot{\theta}) - (-mgl\sin\theta) = 0 \tag{3.36}$$

$$ml^2\ddot{\theta} + mgl\sin\theta = 0 \tag{3.37}$$

$$\ddot{\theta} + \frac{g}{l}\sin\theta = 0 \tag{3.38}$$

Example 3.5 (Particle in Central Force Field). Using polar coordinates (r, θ):

$$T = \frac{1}{2}m(\dot{r}^2 + r^2\dot{\theta}^2) \tag{3.39}$$

$$V = V(r) \tag{3.40}$$

$$L = \frac{1}{2}m(\dot{r}^2 + r^2\dot{\theta}^2) - V(r) \tag{3.41}$$

Two Lagrange equations emerge:

$$r\text{-equation}: \quad m\ddot{r} - mr\dot{\theta}^2 = -\frac{dV}{dr} \tag{3.42}$$

$$\theta\text{-equation}: \quad \frac{d}{dt}(mr^2\dot{\theta}) = 0 \tag{3.43}$$

The second equation immediately yields conservation of angular momentum: $l = mr^2\dot{\theta} = \text{constant}$.

3.3 Conservative and Non-Conservative Systems

3.3.1 Conservative Forces and Potentials

A force field $\mathbf{F}$ is conservative if:

1. $\nabla \times \mathbf{F} = 0$ (curl-free)
2. The work integral $\int \mathbf{F} \cdot d\mathbf{r}$ is path-independent
3. $\mathbf{F} = -\nabla V$ for some potential function V

For time-independent potentials, energy is conserved:

$$E = T + V = \text{constant} \tag{3.44}$$

This follows directly from Lagrange's equations. Multiply the j-th equation by $\dot{q}_j$ and sum:

$$\sum_j \dot{q}_j \left[\frac{d}{dt}\left(\frac{\partial L}{\partial \dot{q}_j}\right) - \frac{\partial L}{\partial q_j} \right] = 0 \tag{3.45}$$

After manipulation:

$$\frac{dE}{dt} = -\frac{\partial L}{\partial t} \tag{3.46}$$

For time-independent Lagrangians, E is conserved.

3.3.2 Generalized Potentials

Some non-conservative forces can be derived from generalized potentials. The Lorentz force on a charged particle:

$$\mathbf{F} = q(\mathbf{E} + \mathbf{v} \times \mathbf{B}) \tag{3.47}$$

cannot be derived from an ordinary potential but can be obtained from:

$$L = \frac{1}{2} m \mathbf{v}^2 - q(\phi - \mathbf{v} \cdot \mathbf{A}) \tag{3.48}$$

where ϕ is the scalar potential and $\mathbf{A}$ is the vector potential.

3.3.3 Dissipative Forces

Dissipative forces like friction cannot be derived from potentials. For velocity-proportional damping:

$$\mathbf{F}_{\text{friction}} = -b\mathbf{v} \tag{3.49}$$

Rayleigh introduced a dissipation function:

$$\mathcal{F} = \frac{1}{2} \sum_i b_i \dot{q}_i^2 \tag{3.50}$$

Modified Lagrange equations:

$$\frac{d}{dt}\left(\frac{\partial L}{\partial \dot{q}_j}\right) - \frac{\partial L}{\partial q_j} + \frac{\partial \mathcal{F}}{\partial \dot{q}_j} = 0 \tag{3.51}$$

3.3.4 External Driving Forces

Time-dependent external forces often appear in physical systems. For a driven harmonic oscillator:

$$L = \frac{1}{2}m\dot{x}^2 - \frac{1}{2}kx^2 + xF_0\cos(\omega t) \tag{3.52}$$

The equation of motion:

$$m\ddot{x} + kx = F_0\cos(\omega t) \tag{3.53}$$

3.3.5 Gauge Transformations

The Lagrangian is not unique. Adding a total time derivative of any function of coordinates and time doesn't change the equations of motion:

$$L' = L + \frac{d}{dt}[f(\mathbf{q}, t)] \tag{3.54}$$

This gauge freedom is profound in field theory and connects to conservation laws through Noether's theorem.

3.4 Invariance and Symmetries

3.4.1 Symmetry Transformations

A symmetry of a physical system is a transformation that leaves the equations of motion invariant. In the Lagrangian formalism, we seek transformations of coordinates and time that leave the action stationary.

Consider an infinitesimal transformation:

$$q_i \to q_i' = q_i + \varepsilon\Psi_i(\mathbf{q}, t) \tag{3.55}$$

$$t \to t' = t + \varepsilon\Phi(\mathbf{q}, t) \tag{3.56}$$

where ε is an infinitesimal parameter.

3.4.2 Invariance of the Action

The action transforms as:

$$S' = \int_{t_1'}^{t_2'} L'(\mathbf{q}', \dot{\mathbf{q}}', t')dt' \tag{3.57}$$

For the transformation to be a symmetry, we require $\delta S = S' - S = 0$ to first order in ε.

After careful calculation, the condition for symmetry is:

$$\delta L = \varepsilon \frac{dG}{dt} \tag{3.58}$$

for some function $G(\mathbf{q}, t)$. This means L can change by at most a total derivative.

3.4.3 Types of Symmetries

Spatial Translation Invariance: If the Lagrangian doesn't depend on a particular coordinate q_k (cyclic coordinate):

$$\frac{\partial L}{\partial q_k} = 0 \tag{3.59}$$

The transformation $q_k \to q_k + \varepsilon$ leaves L invariant.

Time Translation Invariance: If $\partial L / \partial t = 0$, the system is invariant under $t \to t + \varepsilon$.

Rotational Invariance: For a system invariant under rotations about an axis, the Lagrangian doesn't change under angular displacements about that axis.

Scaling Symmetry: Some systems exhibit invariance under scaling transformations:

$$\mathbf{r} \to \lambda \mathbf{r} \tag{3.60}$$

$$t \to \lambda^\alpha t \tag{3.61}$$

3.4.4 Continuous Symmetry Groups

Symmetries often form continuous groups (Lie groups):

- Translations: $\mathbb{R}^3$
- Rotations: $SO(3)$
- Galilean transformations
- Lorentz transformations

Each continuous symmetry corresponds to a conserved quantity via Noether's theorem.

3.4.5 Discrete Symmetries

Not all symmetries are continuous:

- Parity: $\mathbf{r} \to -\mathbf{r}$
- Time reversal: $t \to -t$
- Discrete rotations (crystal symmetries)

These don't lead to conservation laws via Noether's theorem but still constrain the system's behavior.

3.5 Noether's Theorem and Conservation Laws

3.5.1 Statement of Noether's Theorem

Emmy Noether's theorem (1915) provides a profound connection between symmetries and conservation laws:

Theorem 3.1 (Noether's Theorem). For every continuous symmetry of the action, there exists a conserved quantity.

More precisely: If the Lagrangian is invariant (up to a total derivative) under a one-parameter family of transformations, then there exists a quantity that is constant along the equations of motion.

3.5.2 Proof of Noether's Theorem

Consider a one-parameter transformation:

$$q_i \to q_i + \varepsilon\Psi_i(\mathbf{q}, t) \tag{3.62}$$

$$t \to t + \varepsilon\Phi(\mathbf{q}, t) \tag{3.63}$$

The variation of the Lagrangian is:

$$\delta L = \sum_i \left(\frac{\partial L}{\partial q_i}\delta q_i + \frac{\partial L}{\partial \dot{q}_i}\delta \dot{q}_i \right) + \frac{\partial L}{\partial t}\delta t \tag{3.64}$$

After extensive calculation using the chain rule and Lagrange's equations:

$$\delta L = \varepsilon \frac{d}{dt}\left[\sum_i \frac{\partial L}{\partial \dot{q}_i}\Psi_i - \left(L - \sum_i \dot{q}_i \frac{\partial L}{\partial \dot{q}_i}\right)\Phi\right] \tag{3.65}$$

If $\delta L = \varepsilon dG/dt$ (symmetry condition), then:

$$\frac{d}{dt}\left[\sum_i \frac{\partial L}{\partial \dot{q}_i}\Psi_i - H\Phi - G\right] = 0 \tag{3.66}$$

where $H = \sum_i \dot{q}_i(\partial L/\partial \dot{q}_i) - L$ is the energy function.

The conserved quantity is:

$$I = \sum_i p_i \Psi_i - H\Phi - G \tag{3.67}$$

where $p_i = \partial L/\partial \dot{q}_i$ are the generalized momenta.

3.5.3 Applications of Noether's Theorem

Example 3.6 (Conservation of Linear Momentum)**.** For translation invariance in the x-direction:

$$\Psi_x = 1, \quad \text{all other } \Psi = 0 \tag{3.68}$$

$$\Phi = 0 \tag{3.69}$$

The conserved quantity is:

$$I = p_x = \frac{\partial L}{\partial \dot{x}} \tag{3.70}$$

Example 3.7 (Conservation of Angular Momentum)**.** For rotation about the z-axis:

$$\delta x = -y\delta\theta, \quad \delta y = x\delta\theta, \quad \delta z = 0 \tag{3.71}$$

$$\Psi_x = -y, \quad \Psi_y = x, \quad \Psi_z = 0 \tag{3.72}$$

$$\Phi = 0 \tag{3.73}$$

The conserved quantity is:

$$I = p_x(-y) + p_y(x) = xp_y - yp_x = L_z \tag{3.74}$$

Example 3.8 (Conservation of Energy). For time translation invariance:

$$\Psi_i = 0 \quad \text{for all } i \tag{3.75}$$
$$\Phi = 1 \tag{3.76}$$

The conserved quantity is:

$$I = -H = -\left(\sum_i \dot{q}_i p_i - L\right) = L - \sum_i \dot{q}_i p_i \tag{3.77}$$

For natural systems where T is homogeneous quadratic in velocities, this equals $-(T + V) = -E$.

3.5.4 Gauge Symmetries and Noether's Second Theorem

Noether actually proved two theorems. The second deals with infinite-dimensional symmetry groups (gauge symmetries):

Theorem 3.2 (Noether's Second Theorem). If the action is invariant under an infinite-dimensional group of transformations, then there exist differential identities among the Euler-Lagrange equations.

This theorem is crucial in field theory, leading to:

- Charge conservation in electromagnetism
- Energy-momentum conservation in general relativity
- Current conservation in gauge theories

3.5.5 Modern Applications and Extensions

Noether's theorem extends far beyond classical mechanics:

Quantum Mechanics: Symmetries become operators, and conserved quantities become good quantum numbers.

Field Theory: Continuous fields have infinite degrees of freedom, leading to conserved currents:

$$\partial_\mu j^\mu = 0 \tag{3.78}$$

Particle Physics: The Standard Model's structure is largely determined by gauge symmetries and their associated conservation laws.

General Relativity: Diffeomorphism invariance leads to the contracted Bianchi identities, expressing local conservation of energy-momentum.

Summary

The Lagrangian formalism transforms mechanics from a study of forces and accelerations to an analysis of scalar energy functions and symmetries. Starting from generalized coordinates that respect constraints naturally, we derived Lagrange's equations from D'Alembert's principle. These equations, emerging from a single scalar function $L = T - V$, automatically incorporate constraint forces and reveal conserved quantities.

The formalism's true power appears in its treatment of symmetries. Noether's theorem establishes a profound connection: every continuous symmetry yields a conservation law. This principle extends throughout physics, from classical mechanics to quantum field theory and general relativity.

The Lagrangian approach offers several advantages:

1. Coordinate independence: equations take the same form in any coordinate system
2. Constraint forces are automatically handled
3. Symmetries and conservation laws are manifest
4. Extension to fields and quantum mechanics is natural
5. The variational principle provides powerful approximation methods

As we'll see in the next chapter, the Hamiltonian formulation provides a complementary perspective, emphasizing phase space and preparing the ground for quantum mechanics. Together, these formulations reveal the deep mathematical structure underlying physical law.

Chapter 4

Hamiltonian Formalism

The Hamiltonian formulation of mechanics, developed by William Rowan Hamilton in the 1830s, provides an alternative to the Lagrangian approach that proves particularly powerful for understanding the geometric structure of mechanics, connecting classical and quantum physics, and solving practical problems in celestial mechanics and dynamical systems. While mathematically equivalent to the Lagrangian formulation, the Hamiltonian perspective emphasizes different aspects: phase space over configuration space, canonical momenta over velocities, and symplectic geometry over variational principles.

This reformulation transforms mechanics into a first-order theory in phase space, where positions and momenta play symmetric roles. This symmetry reveals deep mathematical structures—symplectic geometry, Poisson brackets, and canonical transformations—that extend far beyond classical mechanics into quantum theory, statistical mechanics, and modern mathematical physics. The Hamiltonian approach also provides the natural language for perturbation theory, chaos, and integrable systems.

4.1 Legendre Transformation

4.1.1 Mathematical Foundation of the Legendre Transform

The Legendre transformation is a fundamental mathematical operation that transforms a function of one set of variables into a function of their conjugate variables. In mechanics, it converts the Lagrangian $L(\mathbf{q}, \dot{\mathbf{q}}, t)$, a function of positions and velocities, into the Hamiltonian $H(\mathbf{q}, \mathbf{p}, t)$, a function of positions and momenta.

For a convex function $f(x)$, the Legendre transform is defined as:

$$g(p) = \sup_x [px - f(x)] \tag{4.1}$$

where $p = df/dx$. Geometrically, this transformation replaces a function by the envelope of its tangent lines, encoding the same information in a dual representation.

For mechanics, we start with the Lagrangian $L(\mathbf{q}, \dot{\mathbf{q}}, t)$ and define the canonical momenta:

$$p_i = \frac{\partial L}{\partial \dot{q}_i} \tag{4.2}$$

The Hamiltonian is then:

$$H(\mathbf{q}, \mathbf{p}, t) = \sum_i p_i \dot{q}_i - L(\mathbf{q}, \dot{\mathbf{q}}, t) \tag{4.3}$$

where $\dot{q}_i$ must be expressed in terms of $\mathbf{q}$ and $\mathbf{p}$ by inverting the momentum relations.

4.1.2 Conditions for the Legendre Transform

The Legendre transformation requires certain regularity conditions. The key requirement is that the Hessian matrix:

$$W_{ij} = \frac{\partial^2 L}{\partial \dot{q}_i \partial \dot{q}_j} \tag{4.4}$$

be non-singular ($\det W \neq 0$). This ensures the momentum equations $p_i = \partial L / \partial \dot{q}_i$ can be inverted to express velocities in terms of momenta.

For natural systems where $L = T - V$ with T quadratic in velocities:

$$T = \frac{1}{2} \sum_{ij} M_{ij}(\mathbf{q}) \dot{q}_i \dot{q}_j \tag{4.5}$$

The Hessian equals the mass matrix M_{ij}, which is positive definite for physical systems, guaranteeing invertibility.

Singular Lagrangians: When $\det W = 0$, the system has constraints, and the standard Legendre transform fails. This occurs in gauge theories and requires the Dirac constraint formalism, leading to the distinction between first-class and second-class constraints.

4.1.3 Examples of Legendre Transformations

Example 4.1 (Free Particle).

$$L = \frac{1}{2}m\dot{x}^2 \tag{4.6}$$

$$p = \frac{\partial L}{\partial \dot{x}} = m\dot{x} \quad \Rightarrow \quad \dot{x} = \frac{p}{m} \tag{4.7}$$

$$H = p\dot{x} - L = p\frac{p}{m} - \frac{1}{2}m\left(\frac{p}{m}\right)^2 = \frac{p^2}{2m} \tag{4.8}$$

Example 4.2 (Harmonic Oscillator).

$$L = \frac{1}{2}m\dot{x}^2 - \frac{1}{2}kx^2 \tag{4.9}$$

$$p = m\dot{x} \quad \Rightarrow \quad \dot{x} = \frac{p}{m} \tag{4.10}$$

$$H = \frac{p^2}{2m} + \frac{1}{2}kx^2 \tag{4.11}$$

Example 4.3 (Charged Particle in Electromagnetic Field).

$$L = \frac{1}{2}m\dot{\mathbf{r}}^2 - q(\phi - \dot{\mathbf{r}} \cdot \mathbf{A}) \tag{4.12}$$

$$p_i = \frac{\partial L}{\partial \dot{x}_i} = m\dot{x}_i + qA_i \quad \Rightarrow \quad \dot{x}_i = \frac{p_i - qA_i}{m} \tag{4.13}$$

$$H = \frac{1}{2m}\sum_i (p_i - qA_i)^2 + q\phi \tag{4.14}$$

This example shows how the canonical momentum differs from the mechanical momentum $m\dot{\mathbf{r}}$ by the electromagnetic contribution $q\mathbf{A}$.

4.1.4 Properties of the Legendre Transform

The Legendre transformation has several important properties:

1. **Involution**: Applying the Legendre transform twice returns the original function (up to sign): $\mathcal{L}\mathcal{L}[f] = f$
2. **Convexity**: The transform of a convex function is convex.
3. **Young's Inequality**: For conjugate variables: $xy \leq f(x) + g(y)$ with equality when $y = f'(x)$.
4. **Scaling**: If $f(\lambda x) = \lambda^\alpha f(x)$, then $g(\lambda p) = \lambda^\beta g(p)$ where $\alpha + \beta = \alpha\beta$.

4.1.5 Physical Interpretation

The Legendre transformation exchanges the role of "slopes" and "intercepts." In the Lagrangian formulation, velocities are fundamental and momenta are derived. The Hamiltonian formulation reverses this: momenta become fundamental while velocities are derived as:

$$\dot{q}_i = \frac{\partial H}{\partial p_i} \tag{4.15}$$

This duality reflects a deep symmetry in mechanics between configuration and momentum representations, foreshadowing the position-momentum duality in quantum mechanics.

4.2 Canonical Equations

4.2.1 Derivation of Hamilton's Equations

Starting from the definition $H = \sum_i p_i \dot{q}_i - L$, we compute the differential:

$$dH = \sum_i (p_i d\dot{q}_i + \dot{q}_i dp_i) - \frac{\partial L}{\partial q_i} dq_i - \frac{\partial L}{\partial \dot{q}_i} d\dot{q}_i - \frac{\partial L}{\partial t} dt \tag{4.16}$$

Using $p_i = \partial L / \partial \dot{q}_i$ and the Euler-Lagrange equations $\partial L / \partial q_i = d/dt(\partial L / \partial \dot{q}_i) = \dot{p}_i$:

$$dH = \sum_i (\dot{q}_i dp_i - \dot{p}_i dq_i) - \frac{\partial L}{\partial t} dt \tag{4.17}$$

Since $H = H(\mathbf{q}, \mathbf{p}, t)$, we also have:

$$dH = \sum_i \left(\frac{\partial H}{\partial q_i} dq_i + \frac{\partial H}{\partial p_i} dp_i \right) + \frac{\partial H}{\partial t} dt \tag{4.18}$$

Comparing coefficients yields Hamilton's canonical equations:

$$\dot{q}_i = \frac{\partial H}{\partial p_i} \tag{4.19}$$

$$\dot{p}_i = -\frac{\partial H}{\partial q_i} \tag{4.20}$$

These $2n$ first-order equations replace the n second-order Lagrange equations.

4.2.2 Structure and Symmetry

Hamilton's equations exhibit remarkable symmetry between $\mathbf{q}$ and $\mathbf{p}$. This symmetry suggests that phase space, not configuration space, is the natural arena for mechanics. The equations can be written in matrix form:

$$\begin{pmatrix} \dot{\mathbf{q}} \\ \dot{\mathbf{p}} \end{pmatrix} = J \begin{pmatrix} \frac{\partial H}{\partial \mathbf{q}} \\ \frac{\partial H}{\partial \mathbf{p}} \end{pmatrix} \tag{4.21}$$

where J is the symplectic matrix:

$$J = \begin{pmatrix} 0 & I \\ -I & 0 \end{pmatrix} \tag{4.22}$$

This formulation reveals that Hamiltonian flow preserves the symplectic structure of phase space.

4.2.3 Conservation of Energy

The time evolution of the Hamiltonian along a trajectory is:

$$\frac{dH}{dt} = \sum_i \left(\frac{\partial H}{\partial q_i} \dot{q}_i + \frac{\partial H}{\partial p_i} \dot{p}_i \right) + \frac{\partial H}{\partial t} \tag{4.23}$$

Substituting Hamilton's equations:

$$\frac{dH}{dt} = \sum_i \left(\frac{\partial H}{\partial q_i} \frac{\partial H}{\partial p_i} - \frac{\partial H}{\partial p_i} \frac{\partial H}{\partial q_i} \right) + \frac{\partial H}{\partial t} = \frac{\partial H}{\partial t} \tag{4.24}$$

Therefore, H is conserved if and only if $\partial H/\partial t = 0$ (time-independent Hamiltonian).

4.2.4 Examples and Applications

Example 4.4 (Kepler Problem). In polar coordinates (r, θ) with momenta (p_r, p_θ):

$$H = \frac{p_r^2}{2m} + \frac{p_\theta^2}{2mr^2} - \frac{k}{r} \tag{4.25}$$

Hamilton's equations:

$$\dot{r} = \frac{\partial H}{\partial p_r} = \frac{p_r}{m} \tag{4.26}$$

$$\dot{\theta} = \frac{\partial H}{\partial p_\theta} = \frac{p_\theta}{mr^2} \tag{4.27}$$

$$\dot{p}_r = -\frac{\partial H}{\partial r} = \frac{p_\theta^2}{mr^3} - \frac{k}{r^2} \tag{4.28}$$

$$\dot{p}_\theta = -\frac{\partial H}{\partial \theta} = 0 \tag{4.29}$$

The last equation shows angular momentum p_θ is conserved, reducing the problem to one dimension.

Example 4.5 (Coupled Oscillators)**.** For two masses connected by springs:

$$H = \frac{p_1^2 + p_2^2}{2m} + \frac{1}{2}k_1 x_1^2 + \frac{1}{2}k_2(x_2 - x_1)^2 + \frac{1}{2}k_3 x_2^2 \tag{4.30}$$

The equations reveal normal modes through linear combinations of coordinates.

4.2.5 Hamilton's Equations from Variational Principle

Hamilton's equations also follow from a modified variational principle. The action in phase space is:

$$S = \int \left[\sum_i p_i \dot{q}_i - H(\mathbf{q}, \mathbf{p}, t) \right] dt \tag{4.31}$$

Varying with respect to $\mathbf{q}$ and $\mathbf{p}$ independently yields Hamilton's equations. This phase space action principle proves crucial in path integral quantization.

4.3 Phase Space

4.3.1 Geometry of Phase Space

Phase space is the $2n$-dimensional space with coordinates $(q_1, \ldots, q_n, p_1, \ldots, p_n)$. Each point represents a complete dynamical state, and time evolution traces curves in this space. Unlike configuration space, phase space treats positions and momenta symmetrically.

The fundamental geometric structure is the symplectic form:

$$\omega = \sum_i dq_i \wedge dp_i \tag{4.32}$$

This defines a non-degenerate, antisymmetric bilinear form that Hamiltonian evolution preserves.

4.3.2 Liouville's Theorem

One of the most profound results in Hamiltonian mechanics is Liouville's theorem:

Theorem 4.1 (Liouville's Theorem)**.** The phase space volume element $d\Gamma = \prod_i dq_i dp_i$ is preserved under Hamiltonian evolution.

Proof. Consider a volume element evolving under the flow. After time dt:

$$q_i(t + dt) = q_i(t) + \dot{q}_i dt = q_i + \frac{\partial H}{\partial p_i} dt \tag{4.33}$$

$$p_i(t + dt) = p_i(t) + \dot{p}_i dt = p_i - \frac{\partial H}{\partial q_i} dt \tag{4.34}$$

The Jacobian of this transformation is:

$$J = \det \left| \frac{\partial(\mathbf{q}', \mathbf{p}')}{\partial(\mathbf{q}, \mathbf{p})} \right| = \det \begin{pmatrix} I + dt \frac{\partial^2 H}{\partial \mathbf{q} \partial \mathbf{p}} & dt \frac{\partial^2 H}{\partial \mathbf{p}^2} \\ -dt \frac{\partial^2 H}{\partial \mathbf{q}^2} & I - dt \frac{\partial^2 H}{\partial \mathbf{q} \partial \mathbf{p}} \end{pmatrix} \tag{4.35}$$

To first order in dt:

$$J = 1 + dt \cdot \mathrm{Tr} \left(\frac{\partial^2 H}{\partial \mathbf{q} \partial \mathbf{p}} - \frac{\partial^2 H}{\partial \mathbf{p} \partial \mathbf{q}} \right) = 1 \tag{4.36}$$

Therefore, volume is preserved. □

4.3.3 Phase Portraits and Fixed Points

Phase space analysis provides powerful visualization tools:

Fixed Points: Solutions where $\dot{\mathbf{q}} = \dot{\mathbf{p}} = 0$, corresponding to equilibria.

Classification: Near a fixed point, linearization yields:

$$\delta\dot{\mathbf{x}} = Df \cdot \delta\mathbf{x} \tag{4.37}$$

where Df is the Jacobian. Eigenvalues determine stability:

- Real, opposite signs: saddle point (unstable)
- Real, same sign: node (stable if negative)
- Complex conjugate: focus or center

Example 4.6 (Pendulum Phase Portrait).

$$H = \frac{p^2}{2ml^2} - mgl\cos\theta \tag{4.38}$$

Fixed points: $(\theta, p) = (0, 0)$ [stable center] and $(\pi, 0)$ [unstable saddle]. The separatrix connecting saddle points divides oscillatory and rotational motion.

4.3.4 Poincaré Sections

For systems with more than one degree of freedom, full phase space visualization becomes impossible. Poincaré sections reduce dimensionality by recording intersections of trajectories with a lower-dimensional surface.

For a 2-DOF system with energy conservation, the 3D energy surface can be sectioned by a plane (e.g., $q_2 = 0$ with $p_2 > 0$). Regular trajectories produce curves, while chaotic trajectories fill areas.

4.3.5 Action-Angle Variables

For integrable systems, special canonical coordinates (I, θ) called action-angle variables exist where:

- Actions I_i are constants of motion
- Angles θ_i increase linearly with time: $\theta_i = \omega_i(I)t + \theta_i^0$
- H depends only on actions: $H = H(I)$

Actions are adiabatic invariants, remaining approximately constant under slow parameter changes—crucial for plasma physics and accelerator design.

4.4 Poisson Brackets

4.4.1 Definition and Properties

The Poisson bracket of two phase space functions $f(\mathbf{q}, \mathbf{p}, t)$ and $g(\mathbf{q}, \mathbf{p}, t)$ is:

$$\{f, g\} = \sum_i \left(\frac{\partial f}{\partial q_i} \frac{\partial g}{\partial p_i} - \frac{\partial f}{\partial p_i} \frac{\partial g}{\partial q_i} \right) \tag{4.39}$$

This operation encodes the symplectic structure algebraically.
Fundamental Poisson Brackets:

$$\{q_i, q_j\} = 0 \tag{4.40}$$
$$\{p_i, p_j\} = 0 \tag{4.41}$$
$$\{q_i, p_j\} = \delta_{ij} \tag{4.42}$$

Properties:

1. Antisymmetry: $\{f, g\} = -\{g, f\}$
2. Linearity: $\{af + bg, h\} = a\{f, h\} + b\{g, h\}$
3. Leibniz rule: $\{fg, h\} = f\{g, h\} + \{f, h\}g$
4. Jacobi identity: $\{f, \{g, h\}\} + \{g, \{h, f\}\} + \{h, \{f, g\}\} = 0$

4.4.2 Time Evolution and Constants of Motion

The time evolution of any phase space function $f(\mathbf{q}, \mathbf{p}, t)$ is:

$$\frac{df}{dt} = \sum_i \left(\frac{\partial f}{\partial q_i} \dot{q}_i + \frac{\partial f}{\partial p_i} \dot{p}_i \right) + \frac{\partial f}{\partial t} \tag{4.43}$$

Using Hamilton's equations:

$$\frac{df}{dt} = \{f, H\} + \frac{\partial f}{\partial t} \tag{4.44}$$

This fundamental equation shows:

- f is conserved if $\{f, H\} + \partial f / \partial t = 0$
- For time-independent f, conservation requires $\{f, H\} = 0$

4.4.3 Poisson's Theorem

Theorem 4.2 (Poisson's Theorem). If f and g are constants of motion, then $\{f, g\}$ is also a constant of motion.

Proof. Using Jacobi's identity:

$$\frac{d\{f,g\}}{dt} = \{\{f,g\},H\} = -\{f,\{g,H\}\} - \{g,\{H,f\}\} = 0 \tag{4.45}$$

□

This provides a method for generating new conserved quantities, though the result may be trivial (constant) or functionally dependent on known integrals.

4.4.4 Examples and Applications

Example 4.7 (Angular Momentum Components). For $L_x = yp_z - zp_y$, $L_y = zp_x - xp_z$, $L_z = xp_y - yp_x$:

$$\{L_x, L_y\} = L_z \quad \text{(and cyclic permutations)} \tag{4.46}$$

These relations encode the $SO(3)$ Lie algebra structure.

Example 4.8 (Harmonic Oscillator Ladder Operators). Define $a = (m\omega q + ip)/\sqrt{2m\omega}$, $a^* = (m\omega q - ip)/\sqrt{2m\omega}$:

$$\{a, a^*\} = -i \tag{4.47}$$

$$H = \omega\left(a^*a + \frac{1}{2}\right) \tag{4.48}$$

This anticipates quantum creation/annihilation operators.

4.4.5 Connection to Quantum Mechanics

The Poisson bracket structure translates directly to quantum mechanics through:

$$\{f, g\} \to \frac{1}{i\hbar}[\hat{f}, \hat{g}] \tag{4.49}$$

This correspondence principle guided the development of quantum mechanics, with:

- Classical observables → Hermitian operators
- Poisson brackets → Commutators
- Classical evolution → Heisenberg equation

4.5 Theory of Canonical Transformations

4.5.1 Definition and Motivation

A canonical transformation is a change of phase space coordinates:

$$(\mathbf{q}, \mathbf{p}) \to (\mathbf{Q}, \mathbf{P}) \tag{4.50}$$

that preserves the form of Hamilton's equations. That is, if $(\mathbf{q}, \mathbf{p})$ satisfy Hamilton's equations with Hamiltonian H, then $(\mathbf{Q}, \mathbf{P})$ satisfy:

$$\dot{Q}_i = \frac{\partial K}{\partial P_i} \tag{4.51}$$

$$\dot{P}_i = -\frac{\partial K}{\partial Q_i} \tag{4.52}$$

for some new Hamiltonian $K(\mathbf{Q}, \mathbf{P}, t)$.

Canonical transformations are the "symmetries" of Hamiltonian mechanics, preserving its mathematical structure while potentially simplifying specific problems.

4.5.2 Generating Functions

Canonical transformations can be characterized by generating functions. There are four types, related by Legendre transforms:

Type 1: $F_1(\mathbf{q}, \mathbf{Q}, t)$

$$p_i = \frac{\partial F_1}{\partial q_i}, \quad P_i = -\frac{\partial F_1}{\partial Q_i} \tag{4.53}$$

$$K = H + \frac{\partial F_1}{\partial t} \tag{4.54}$$

Type 2: $F_2(\mathbf{q}, \mathbf{P}, t) = F_1 + \sum_i Q_i P_i$

$$p_i = \frac{\partial F_2}{\partial q_i}, \quad Q_i = \frac{\partial F_2}{\partial P_i} \tag{4.55}$$

$$K = H + \frac{\partial F_2}{\partial t} \tag{4.56}$$

Type 3: $F_3(\mathbf{p}, \mathbf{Q}, t) = F_1 - \sum_i q_i p_i$

$$q_i = -\frac{\partial F_3}{\partial p_i}, \quad P_i = -\frac{\partial F_3}{\partial Q_i} \tag{4.57}$$

$$K = H + \frac{\partial F_3}{\partial t} \tag{4.58}$$

Type 4: $F_4(\mathbf{p}, \mathbf{P}, t) = F_1 + \sum_i Q_i P_i - \sum_i q_i p_i$

$$q_i = -\frac{\partial F_4}{\partial p_i}, \quad Q_i = \frac{\partial F_4}{\partial P_i} \tag{4.59}$$

$$K = H + \frac{\partial F_4}{\partial t} \tag{4.60}$$

4.5.3 Examples of Canonical Transformations

Example 4.9 (Point Transformation). $Q_i = Q_i(\mathbf{q}, t)$, with generating function:

$$F_2 = \sum_i Q_i(\mathbf{q}, t) P_i \tag{4.61}$$

This yields $p_i = \sum_j \frac{\partial Q_j}{\partial q_i} P_j$, recovering the momentum transformation law.

Example 4.10 (Exchange of q and p).

$$F_1(\mathbf{q}, \mathbf{Q}) = \sum_i q_i Q_i \tag{4.62}$$

$$p_i = Q_i, \quad P_i = -q_i \tag{4.63}$$

This transformation exchanges positions and momenta, demonstrating their symmetric roles.

Example 4.11 (Harmonic Oscillator to Action-Angle).

$$F_2(q, P) = \frac{m\omega}{2} q^2 \cot P \tag{4.64}$$

$$Q = \frac{\partial F_2}{\partial P} = -\frac{m\omega}{2} \frac{q^2}{\sin^2 P} \tag{4.65}$$

$$p = \frac{\partial F_2}{\partial q} = m\omega q \cot P \tag{4.66}$$

Solving yields action-angle variables with $H = \omega I$.

4.5.4 Symplectic Condition

A transformation is canonical if and only if it preserves the symplectic 2-form:

$$\sum_i dQ_i \wedge dP_i = \sum_i dq_i \wedge dp_i \tag{4.67}$$

In matrix notation, the Jacobian $M = \partial(\mathbf{Q}, \mathbf{P})/\partial(\mathbf{q}, \mathbf{p})$ must satisfy:

$$M^T J M = J \tag{4.68}$$

where J is the symplectic matrix. Such matrices form the symplectic group $Sp(2n)$.

4.5.5 Infinitesimal Canonical Transformations

Consider transformations near the identity:

$$Q_i = q_i + \varepsilon \frac{\partial G}{\partial p_i} \tag{4.69}$$

$$P_i = p_i - \varepsilon \frac{\partial G}{\partial q_i} \tag{4.70}$$

where $G(\mathbf{q}, \mathbf{p})$ is the generator. The change in any function f is:

$$\delta f = \varepsilon \{f, G\} \tag{4.71}$$

This shows:

- Translations: $G = p_x$ generates $q_x \to q_x + \varepsilon$
- Rotations: $G = L_z$ generates rotations about z-axis
- Time evolution: $G = H$ generates time translation

4.5.6 Hamilton-Jacobi Theory

The Hamilton-Jacobi equation seeks a canonical transformation to coordinates where all momenta are constants. If $S(\mathbf{q}, \boldsymbol{\alpha}, t)$ is a complete integral of:

$$\frac{\partial S}{\partial t} + H\left(\mathbf{q}, \frac{\partial S}{\partial \mathbf{q}}, t\right) = 0 \tag{4.72}$$

then the transformation with $F_2 = S(\mathbf{q}, \mathbf{P}, t)$ yields:

$$Q_i = \frac{\partial S}{\partial P_i} = \beta_i \quad \text{(constants)} \tag{4.73}$$

$$K = 0 \quad \text{(all motion is trivial)} \tag{4.74}$$

The solution is:

$$S = \int L\, dt = \text{action} \tag{4.75}$$

connecting to the path integral formulation of quantum mechanics.

4.5.7 Integral Invariants

Canonical transformations preserve certain integral invariants beyond phase space volume:

Poincaré Integral Invariants:

- $\oint \sum_i p_i dq_i$ is preserved for any closed curve
- $\iint \sum_i dp_i \wedge dq_i$ is preserved for any 2-surface

These invariants characterize canonical transformations and connect to topological aspects of phase space.

Summary

The Hamiltonian formulation transforms classical mechanics into a first-order theory on phase space, revealing profound mathematical structures. The Legendre transformation exchanges the velocity-based Lagrangian description for a momentum-based approach, yielding Hamilton's canonical equations with their striking q-p symmetry.

Phase space emerges as the natural arena for mechanics, with Liouville's theorem guaranteeing volume preservation—a result crucial for statistical mechanics. The Poisson bracket provides an algebraic structure encoding dynamics, with time evolution generated by bracketing with the Hamiltonian. This structure maps directly to quantum mechanical commutators.

Canonical transformations, the "symmetries" of phase space, preserve the form of Hamilton's equations while potentially simplifying problems. Generated by functions satisfying the symplectic condition, they include point transformations, momentum-position exchanges, and the profound Hamilton-Jacobi transformation that reduces dynamics to statics.

The Hamiltonian framework extends throughout physics:

- Quantum mechanics emerges via canonical quantization
- Statistical mechanics builds on phase space and Liouville's theorem
- Chaos theory uses phase portraits and Poincaré sections
- Symplectic integrators preserve geometric structure numerically
- Modern physics, from gauge theories to string theory, uses Hamiltonian methods

Together with the Lagrangian formulation, the Hamiltonian approach provides complementary perspectives on the principle of least action, revealing the deep mathematical beauty underlying physical law.

Part II

Applications in Classical Physics

Part II demonstrates the power and versatility of the principle of least action by applying it to increasingly complex classical systems. We begin with point particle mechanics, progress through rigid bodies and continua, and culminate with classical field theory. Throughout, we'll see how the variational formulation not only simplifies calculations but also reveals deep physical insights and symmetries that would be obscure in the Newtonian approach.

The applications in this part serve multiple purposes: they provide concrete examples of the abstract formalism developed in Part I, they introduce techniques that extend to quantum and relativistic physics, and they solve problems of practical importance in physics and engineering. We'll particularly emphasize how conservation laws emerge naturally from symmetries and how the choice of coordinates can dramatically simplify complex problems.

Chapter 5

Point Particle Mechanics

Point particle mechanics forms the foundation of classical dynamics. Despite the idealization of treating objects as point masses, this approximation captures essential physics in countless situations—from planetary orbits to molecular vibrations. The Lagrangian and Hamiltonian approaches transform these problems, revealing hidden symmetries and providing systematic solution methods that extend far beyond what's practical with Newton's laws alone.

This chapter explores four fundamental classes of problems: motion in central fields (including the famous Kepler problem), the two-body problem and its reduction to single-particle motion, coupled oscillator systems that model everything from molecules to coupled pendulums, and the general N-body problem that, despite its analytical intractability, yields crucial insights into celestial mechanics and statistical physics.

5.1 Motion in a Central Field

5.1.1 General Formulation

A central force depends only on the distance from a fixed center and points along the radial direction:

$$\mathbf{F} = F(r)\hat{\mathbf{r}} \tag{5.1}$$

where r is the distance from the origin and $\hat{\mathbf{r}}$ is the unit radial vector. Such forces arise from fundamental interactions—gravitational, electrostatic, and nuclear forces all have this character for spherically symmetric sources.

The potential energy for a central force is:

$$V(r) = -\int F(r)\,dr \tag{5.2}$$

Common examples include:

- Gravitational/Coulomb: $V(r) = -k/r$ (attractive for $k > 0$)
- Harmonic oscillator: $V(r) = \frac{1}{2}kr^2$
- Yukawa potential: $V(r) = -\frac{g^2}{r}e^{-\mu r}$
- Lennard-Jones: $V(r) = 4\varepsilon\left[\left(\frac{\sigma}{r}\right)^{12} - \left(\frac{\sigma}{r}\right)^{6}\right]$

5.1.2 Lagrangian in Spherical Coordinates

The natural coordinates for central force problems are spherical coordinates (r, θ, ϕ). The position vector is:

$$\mathbf{r} = r\hat{\mathbf{r}} = r(\sin\theta\cos\phi\,\hat{\mathbf{x}} + \sin\theta\sin\phi\,\hat{\mathbf{y}} + \cos\theta\,\hat{\mathbf{z}}) \tag{5.3}$$

The velocity components are:

$$\dot{\mathbf{r}} = \dot{r}\hat{\mathbf{r}} + r\dot{\theta}\hat{\boldsymbol{\theta}} + r\sin\theta\dot{\phi}\hat{\boldsymbol{\phi}} \tag{5.4}$$

where $\hat{\boldsymbol{\theta}}$ and $\hat{\boldsymbol{\phi}}$ are unit vectors in the θ and ϕ directions.

The kinetic energy becomes:

$$T = \frac{1}{2}m|\dot{\mathbf{r}}|^2 = \frac{1}{2}m(\dot{r}^2 + r^2\dot{\theta}^2 + r^2\sin^2\theta\dot{\phi}^2) \tag{5.5}$$

The Lagrangian is:

$$L = T - V = \frac{1}{2}m(\dot{r}^2 + r^2\dot{\theta}^2 + r^2\sin^2\theta\dot{\phi}^2) - V(r) \tag{5.6}$$

5.1.3 Conservation Laws from Symmetries

The central force problem exhibits several symmetries:

Rotational Symmetry: The Lagrangian is independent of ϕ (cyclic coordinate), yielding:

$$p_\phi = \frac{\partial L}{\partial\dot{\phi}} = mr^2\sin^2\theta\dot{\phi} = L_z = \text{constant} \tag{5.7}$$

This is the z-component of angular momentum.

Planar Motion: Since angular momentum $\mathbf{L} = \mathbf{r} \times m\dot{\mathbf{r}}$ is conserved and $\mathbf{r} \cdot \mathbf{L} = 0$, the motion remains in a plane perpendicular to $\mathbf{L}$. Choosing coordinates so $\mathbf{L}$ points along the z-axis, we have $\theta = \pi/2$ and $\dot{\theta} = 0$.

Effective One-Dimensional Problem: In the orbital plane with $\theta = \pi/2$:

$$L = \frac{1}{2}m(\dot{r}^2 + r^2\dot{\phi}^2) - V(r) \tag{5.8}$$

Using $L_z = mr^2\dot{\phi} = l$ (constant):

$$L = \frac{1}{2}m\dot{r}^2 + \frac{l^2}{2mr^2} - V(r) \tag{5.9}$$

This describes one-dimensional motion in an effective potential:

$$V_{\text{eff}}(r) = V(r) + \frac{l^2}{2mr^2} \tag{5.10}$$

5.1.4 Equations of Motion and Orbits

The radial equation from Lagrange's equations:

$$m\ddot{r} = -\frac{dV_{\text{eff}}}{dr} = -\frac{dV}{dr} + \frac{l^2}{mr^3} \tag{5.11}$$

Energy conservation gives:

$$E = \frac{1}{2}m\dot{r}^2 + V_{\text{eff}}(r) \tag{5.12}$$

Solving for $\dot{r}$:

$$\dot{r} = \pm\sqrt{\frac{2}{m}[E - V_{\text{eff}}(r)]} \tag{5.13}$$

The orbit equation is found by eliminating time:

$$d\phi = \frac{l}{mr^2}dt = \frac{l}{mr^2}\frac{dr}{\dot{r}} \tag{5.14}$$

Integrating:

$$\phi - \phi_0 = \int \frac{l/r^2\, dr}{\sqrt{2m[E - V(r)] - l^2/r^2}} \tag{5.15}$$

This integral determines $r(\phi)$, the orbital shape.

5.1.5 The Kepler Problem

For the gravitational/Coulomb potential $V(r) = -k/r$:

The effective potential is:

$$V_{\text{eff}}(r) = -\frac{k}{r} + \frac{l^2}{2mr^2} \tag{5.16}$$

This has a minimum at $r_0 = l^2/mk$ with $V_{\text{eff}}(r_0) = -mk^2/2l^2$.

Orbital Equation: The integral for ϕ yields:

$$r = \frac{p}{1 + e\cos(\phi - \phi_0)} \tag{5.17}$$

where:

$$p = \frac{l^2}{mk} \quad \text{(semi-latus rectum)} \tag{5.18}$$

$$e = \sqrt{1 + \frac{2El^2}{mk^2}} \quad \text{(eccentricity)} \tag{5.19}$$

Classification of Orbits:

- $E < 0$, $e < 1$: Ellipse (bound orbit)
- $E = 0$, $e = 1$: Parabola (marginally bound)
- $E > 0$, $e > 1$: Hyperbola (unbound)
- $E = -mk^2/2l^2$, $e = 0$: Circle

Kepler's Laws:

1. Orbits are conic sections with the force center at a focus
2. Equal areas swept in equal times (conservation of angular momentum)
3. $T^2 \propto a^3$ for elliptical orbits (period-semimajor axis relation)

5.1.6 Additional Symmetries and the Runge-Lenz Vector

The Kepler problem possesses a hidden symmetry. The Runge-Lenz vector:

$$\mathbf{A} = \mathbf{p} \times \mathbf{L} - mk\hat{\mathbf{r}} \tag{5.20}$$

is conserved. In components:

$$A_i = \varepsilon_{ijk} p_j L_k - mk\frac{x_i}{r} \tag{5.21}$$

This vector points toward perihelion and has magnitude mke.

The Poisson brackets of $\mathbf{L}$ and $\mathbf{A}$ form the $so(4)$ algebra for $E < 0$, explaining the "accidental" degeneracy of hydrogen energy levels and the closure of elliptical orbits.

5.2 Two-Body Problem

5.2.1 General Formulation

Consider two particles with masses m_1 and m_2 at positions $\mathbf{r}_1$ and $\mathbf{r}_2$, interacting through a potential $V(|\mathbf{r}_1 - \mathbf{r}_2|)$. The Lagrangian is:

$$L = \frac{1}{2}m_1\dot{\mathbf{r}}_1^2 + \frac{1}{2}m_2\dot{\mathbf{r}}_2^2 - V(|\mathbf{r}_1 - \mathbf{r}_2|) \tag{5.22}$$

This system has six degrees of freedom, but symmetries allow significant simplification.

5.2.2 Center of Mass and Relative Coordinates

Introduce center of mass and relative coordinates:

$$\mathbf{R} = \frac{m_1\mathbf{r}_1 + m_2\mathbf{r}_2}{m_1 + m_2} \tag{5.23}$$

$$\mathbf{r} = \mathbf{r}_1 - \mathbf{r}_2 \tag{5.24}$$

The inverse transformation:

$$\mathbf{r}_1 = \mathbf{R} + \frac{m_2}{m_1 + m_2}\mathbf{r} \tag{5.25}$$

$$\mathbf{r}_2 = \mathbf{R} - \frac{m_1}{m_1 + m_2}\mathbf{r} \tag{5.26}$$

5.2.3 Separation of Motion

Substituting into the kinetic energy:

$$T = \frac{1}{2}m_1\dot{\mathbf{r}}_1^2 + \frac{1}{2}m_2\dot{\mathbf{r}}_2^2 \tag{5.27}$$
$$= \frac{1}{2}M\dot{\mathbf{R}}^2 + \frac{1}{2}\mu\dot{\mathbf{r}}^2 \tag{5.28}$$

where:

$$M = m_1 + m_2 \quad \text{(total mass)} \tag{5.29}$$
$$\mu = \frac{m_1 m_2}{m_1 + m_2} \quad \text{(reduced mass)} \tag{5.30}$$

The Lagrangian separates:

$$L = L_{\text{cm}} + L_{\text{rel}} \tag{5.31}$$
$$L_{\text{cm}} = \frac{1}{2}M\dot{\mathbf{R}}^2 \tag{5.32}$$
$$L_{\text{rel}} = \frac{1}{2}\mu\dot{\mathbf{r}}^2 - V(r) \tag{5.33}$$

5.2.4 Conservation Laws

Center of Mass Motion: Since L_{cm} depends only on $\dot{\mathbf{R}}$:

$$\frac{d}{dt}\left(\frac{\partial L_{\text{cm}}}{\partial \dot{\mathbf{R}}}\right) = 0 \quad \Rightarrow \quad \mathbf{P} = M\dot{\mathbf{R}} = \text{constant} \tag{5.34}$$

The center of mass moves with constant velocity.

Relative Motion: The relative motion Lagrangian L_{rel} describes a single particle of mass μ in the central potential $V(r)$. All results from Section 5.1 apply with $m \to \mu$.

Total Angular Momentum:

$$\mathbf{L}_{\text{total}} = \mathbf{L}_{\text{cm}} + \mathbf{L}_{\text{rel}} = \mathbf{R} \times \mathbf{P} + \mathbf{r} \times \mu\dot{\mathbf{r}} \tag{5.35}$$

Both terms are separately conserved for an isolated system.

5.2.5 Energy Relations

The total energy separates:

$$E_{\text{total}} = E_{\text{cm}} + E_{\text{rel}} \tag{5.36}$$

$$E_{\text{cm}} = \frac{\mathbf{P}^2}{2M} \tag{5.37}$$

$$E_{\text{rel}} = \frac{1}{2}\mu\dot{\mathbf{r}}^2 + V(r) \tag{5.38}$$

For bound states ($E_{\text{rel}} < 0$), the system's internal energy is:

$$E_{\text{internal}} = E_{\text{rel}} = E_{\text{total}} - \frac{\mathbf{P}^2}{2M} \tag{5.39}$$

5.2.6 Examples and Applications

Binary Stars: Two stars orbiting their common center of mass. Observable quantities:

- Period T from Kepler's third law with $m \to \mu$
- Individual stellar velocities: $v_1 = \frac{m_2}{M}v_{\text{rel}}$, $v_2 = \frac{m_1}{M}v_{\text{rel}}$
- Mass determination from spectroscopic measurements

Positronium: Electron-positron bound state with $\mu = m_e/2$. Energy levels are half those of hydrogen due to the reduced mass factor.

Molecular Vibrations: For small oscillations about equilibrium separation r_0:

$$V(r) \approx V(r_0) + \frac{1}{2}k(r - r_0)^2 \tag{5.40}$$

Leading to harmonic oscillator with frequency $\omega = \sqrt{k/\mu}$.

5.3 Coupled Oscillators

5.3.1 General Formulation

Consider N particles connected by springs, representing systems from coupled pendulums to crystal lattices. For particles constrained to one dimension with positions x_i and masses m_i:

$$T = \frac{1}{2}\sum_i m_i\dot{x}_i^2 \tag{5.41}$$

For nearest-neighbor coupling with spring constants k_{ij}:

$$V = \frac{1}{2}\sum_{\{i,j\}} k_{ij}(x_i - x_j)^2 \tag{5.42}$$

5.3.2 Small Oscillations and Normal Modes

For small displacements from equilibrium positions x_i^0, let $\xi_i = x_i - x_i^0$. Expanding to quadratic order:

$$T = \frac{1}{2}\sum_i m_i\dot{\xi}_i^2 \tag{5.43}$$

$$V = \frac{1}{2}\sum_{i,j} V_{ij}\xi_i\xi_j \tag{5.44}$$

where $V_{ij} = \left.\frac{\partial^2 V}{\partial x_i \partial x_j}\right|_{\text{equilibrium}}$.

The Lagrangian in matrix form:

$$L = \frac{1}{2}\dot{\boldsymbol{\xi}}^T M\dot{\boldsymbol{\xi}} - \frac{1}{2}\boldsymbol{\xi}^T V\boldsymbol{\xi} \tag{5.45}$$

where M is the mass matrix and V the potential matrix.

5.3.3 Normal Mode Analysis

Seeking solutions $\boldsymbol{\xi}(t) = \mathbf{A}e^{i\omega t}$, the equations of motion yield:

$$(V - \omega^2 M)\mathbf{A} = 0 \tag{5.46}$$

This eigenvalue problem has N solutions $(\omega_\alpha^2, \mathbf{A}_\alpha)$, $\alpha = 1, \ldots, N$.
Orthogonality: The eigenvectors satisfy:

$$\mathbf{A}_\alpha^T M\mathbf{A}_\beta = \delta_{\alpha\beta} \tag{5.47}$$

$$\mathbf{A}_\alpha^T V\mathbf{A}_\beta = \omega_\alpha^2\delta_{\alpha\beta} \tag{5.48}$$

Normal Coordinates: Define $Q_\alpha = \mathbf{A}_\alpha^T M\boldsymbol{\xi}$. The Lagrangian becomes:

$$L = \frac{1}{2}\sum_\alpha(\dot{Q}_\alpha^2 - \omega_\alpha^2 Q_\alpha^2) \tag{5.49}$$

Each normal mode is an independent harmonic oscillator.

5.3.4 Example: Two Coupled Pendulums

Two identical pendulums of mass m and length l coupled by a spring of constant k at distance a from the pivot:

$$L = \frac{1}{2}ml^2(\dot{\theta}_1^2 + \dot{\theta}_2^2) + mgl(\cos\theta_1 + \cos\theta_2) - \frac{1}{2}ka^2(\theta_1 - \theta_2)^2 \tag{5.50}$$

For small angles:

$$T = \frac{1}{2}ml^2(\dot{\theta}_1^2 + \dot{\theta}_2^2) \tag{5.51}$$

$$V = -\frac{1}{2}mgl(\theta_1^2 + \theta_2^2) + \frac{1}{2}ka^2(\theta_1 - \theta_2)^2 \tag{5.52}$$

Normal modes:

- Symmetric: $\theta_1 = \theta_2$, $\omega_1^2 = g/l$
- Antisymmetric: $\theta_1 = -\theta_2$, $\omega_2^2 = g/l + 2ka^2/ml^2$

5.3.5 Continuum Limit and Wave Equations

For $N \to \infty$ with fixed total length L, the discrete index $i \to$ continuous position x:

$$\xi_i(t) \to \psi(x,t) \tag{5.53}$$

$$\sum_i \to \int dx \tag{5.54}$$

The Lagrangian density becomes:

$$\mathcal{L} = \frac{1}{2}\rho\left(\frac{\partial\psi}{\partial t}\right)^2 - \frac{1}{2}\tau\left(\frac{\partial\psi}{\partial x}\right)^2 \tag{5.55}$$

Leading to the wave equation:

$$\frac{\partial^2\psi}{\partial t^2} = \frac{\tau}{\rho}\frac{\partial^2\psi}{\partial x^2} \tag{5.56}$$

5.3.6 Energy Exchange and Beats

For weakly coupled oscillators ($ka^2/mgl \ll 1$), starting with energy in one pendulum leads to slow energy exchange with period:

$$T_{\text{beat}} = \frac{2\pi}{\omega_2 - \omega_1} \approx \frac{2\pi l}{ka^2}\sqrt{\frac{l}{g}} \tag{5.57}$$

This illustrates the general phenomenon of energy sharing in weakly coupled systems.

5.4 N-Body Problem

5.4.1 General Formulation

For N particles with masses m_i at positions $\mathbf{r}_i$ interacting through pair potentials:

$$L = T - V = \frac{1}{2}\sum_i m_i \dot{\mathbf{r}}_i^2 - \sum_{i<j} V_{ij}(|\mathbf{r}_i - \mathbf{r}_j|) \tag{5.58}$$

This system has $3N$ degrees of freedom, but conserved quantities reduce the effective dimensionality.

5.4.2 Conservation Laws and Integrals of Motion

Momentum Conservation: Translation invariance yields:

$$\mathbf{P} = \sum_i m_i \dot{\mathbf{r}}_i = \text{constant} \tag{5.59}$$

Angular Momentum Conservation: Rotational invariance gives:

$$\mathbf{L} = \sum_i \mathbf{r}_i \times m_i \dot{\mathbf{r}}_i = \text{constant} \tag{5.60}$$

Energy Conservation: Time translation invariance ensures:

$$E = T + V = \text{constant} \tag{5.61}$$

These provide 10 integrals of motion (3 + 3 + 1 + 3 for center of mass position).

5.4.3 The Three-Body Problem

Even for $N = 3$, the problem is generally non-integrable. After removing center of mass motion, we have 6 degrees of freedom with only 4 remaining integrals (energy and angular momentum).

Restricted Three-Body Problem: If $m_3 \ll m_1, m_2$, the motion of the third body in the field of two massive bodies in circular orbits yields:

$$L = \frac{1}{2}m_3(\dot{x}^2 + \dot{y}^2) + m_3\omega(x\dot{y} - y\dot{x}) - V(x, y) \tag{5.62}$$

where (x, y) are coordinates in the rotating frame.

Lagrange Points: Five equilibrium points exist:

- L_1, L_2, L_3: Collinear points (unstable)
- L_4, L_5: Triangular points (stable for $m_2/m_1 < 0.04$)

5.4.4 Virial Theorem

For power-law potentials $V = kr^n$, time-averaging yields:

$$\langle T \rangle = \frac{n}{2}\langle V \rangle \tag{5.63}$$

For gravity ($n = -1$): $\langle T \rangle = -\frac{1}{2}\langle V \rangle$

For harmonic oscillator ($n = 2$): $\langle T \rangle = \langle V \rangle$

This relates average kinetic and potential energies without solving the equations of motion.

5.4.5 Numerical Methods and Chaos

The N-body problem requires numerical integration for $N \geq 3$. Key considerations:

Symplectic Integrators: Preserve phase space volume and energy (approximately) over long times:

- Leapfrog/Verlet methods
- Higher-order symplectic schemes

Chaos and Predictability: Small changes in initial conditions lead to exponentially diverging trajectories, characterized by Lyapunov exponents.

Few-Body Choreographies: Special periodic solutions where all bodies follow the same curve:

- Figure-eight solution for three equal masses
- More complex choreographies for $N > 3$

5.4.6 Statistical Approaches

For large N, statistical methods become powerful:

Virial Equilibrium: For self-gravitating systems:

$$2K + U = 0 \tag{5.64}$$

where K is total kinetic energy and U is total potential energy.

Relaxation Time: Two-body encounters lead to energy exchange on timescale:

$$t_{\text{relax}} \sim \frac{N}{8 \ln N} t_{\text{cross}} \tag{5.65}$$

where t_{cross} is the crossing time.

Phase Transitions: N-body systems can exhibit:

- Gravothermal catastrophe (negative specific heat)
- Core collapse in globular clusters
- Violent relaxation in galaxy formation

Summary

Point particle mechanics demonstrates the power of the Lagrangian formulation in solving complex dynamical problems. The central force problem reveals how symmetries lead to conservation laws and effective one-dimensional motion. The two-body problem's exact solution through separation of variables provides the foundation for understanding binary systems throughout physics.

Coupled oscillators introduce normal mode analysis, a technique that extends from molecular vibrations to quantum field theory. The transition from discrete to continuous systems foreshadows field theory. The N-body problem, while generally intractable analytically, yields crucial insights through conservation laws, special solutions, and statistical approaches.

Key themes emerging from these applications:

1. Symmetries dramatically simplify problems through conserved quantities

2. Coordinate choices can transform complex multi-dimensional problems into simpler ones

3. Small oscillations about equilibrium lead to linear problems with powerful solution methods

4. Even simple systems ($N \geq 3$) can exhibit chaos and require statistical approaches

5. The variational formulation provides systematic approaches unavailable in Newtonian mechanics

These techniques form the foundation for understanding more complex systems—rigid bodies, continua, and fields—that we'll explore in subsequent chapters.

Chapter 6

Rigid Body Mechanics

Rigid body mechanics represents a crucial bridge between point particle dynamics and continuum mechanics. A rigid body—an idealized solid where distances between all points remain constant—exhibits rich dynamical behavior despite this constraint. From spinning tops to satellites, from molecular rotations to galactic dynamics, rigid body motion appears throughout physics at all scales.

The Lagrangian formulation proves particularly powerful for rigid bodies. While the Newtonian approach requires tracking forces and torques at every point, the Lagrangian method needs only the kinetic and potential energies expressed in appropriate coordinates. The constraint of rigidity, which would complicate a force-based analysis, is automatically incorporated through the choice of coordinates. This chapter develops the general theory and explores important special cases that illuminate both terrestrial and celestial mechanics.

6.1 Rigid Body Motion

6.1.1 Degrees of Freedom and Configuration Space

A rigid body is a system of particles with fixed relative positions. If we know the positions of three non-collinear points, all other points are determined. Since each point has three coordinates and there are three distance constraints among three points, a rigid body has:

$$\text{Degrees of freedom} = 3 \times 3 - 3 = 6 \tag{6.1}$$

These six degrees of freedom divide naturally into:

- 3 translational: position of the center of mass
- 3 rotational: orientation of the body

This separation suggests using six coordinates: three for the center of mass position $\mathbf{R}$ and three angles for orientation.

6.1.2 Kinematics of Rigid Body Motion

Consider a rigid body with a body-fixed frame $\{\hat{\mathbf{e}}_1, \hat{\mathbf{e}}_2, \hat{\mathbf{e}}_3\}$ attached at point O. Any point P in the body has position:

$$\mathbf{r} = \mathbf{R} + \boldsymbol{\rho} \tag{6.2}$$

where $\mathbf{R}$ locates the origin O and $\boldsymbol{\rho}$ is the position of P relative to O.

For a rigid body, $|\boldsymbol{\rho}|$ is constant for each point. The velocity is:

$$\mathbf{v} = \dot{\mathbf{R}} + \dot{\boldsymbol{\rho}} \tag{6.3}$$

Since $|\boldsymbol{\rho}|$ is constant, $\dot{\boldsymbol{\rho}}$ represents pure rotation:

$$\dot{\boldsymbol{\rho}} = \boldsymbol{\omega} \times \boldsymbol{\rho} \tag{6.4}$$

where $\boldsymbol{\omega}$ is the angular velocity vector. Thus:

$$\mathbf{v} = \dot{\mathbf{R}} + \boldsymbol{\omega} \times \boldsymbol{\rho} \tag{6.5}$$

This decomposition into translation plus rotation is fundamental to rigid body mechanics.

6.1.3 The Inertia Tensor

The kinetic energy of a rigid body is:

$$T = \frac{1}{2} \int \rho(\mathbf{r}) |\mathbf{v}|^2 d^3r \tag{6.6}$$

where $\rho(\mathbf{r})$ is the mass density. Substituting $\mathbf{v} = \dot{\mathbf{R}} + \boldsymbol{\omega} \times \boldsymbol{\rho}$:

$$T = \frac{1}{2} M |\dot{\mathbf{R}}|^2 + M\dot{\mathbf{R}} \cdot (\boldsymbol{\omega} \times \boldsymbol{\rho}^{\mathrm{cm}}) + \frac{1}{2} \int \rho |\boldsymbol{\omega} \times \boldsymbol{\rho}|^2 d^3r \tag{6.7}$$

where M is total mass and $\boldsymbol{\rho}^{\mathrm{cm}}$ is the center of mass position relative to O.

Choosing O at the center of mass eliminates the middle term:

$$T = T_{\mathrm{trans}} + T_{\mathrm{rot}} \tag{6.8}$$

$$T_{\mathrm{trans}} = \frac{1}{2} M |\dot{\mathbf{R}}|^2 \tag{6.9}$$

$$T_{\mathrm{rot}} = \frac{1}{2} \int \rho |\boldsymbol{\omega} \times \boldsymbol{\rho}|^2 d^3 r \tag{6.10}$$

The rotational kinetic energy can be written as:

$$T_{\mathrm{rot}} = \frac{1}{2} \boldsymbol{\omega} \cdot \mathbf{I} \cdot \boldsymbol{\omega} = \frac{1}{2} \sum_{ij} I_{ij} \omega_i \omega_j \tag{6.11}$$

where the inertia tensor is:

$$I_{ij} = \int \rho(\mathbf{r}) [\delta_{ij} |\boldsymbol{\rho}|^2 - \rho_i \rho_j] d^3 r \tag{6.12}$$

6.1.4 Principal Axes and Moments of Inertia

The inertia tensor is a real, symmetric 3×3 matrix, hence diagonalizable. Its eigenvectors define the principal axes, and eigenvalues I_1, I_2, I_3 are the principal moments of inertia.

In the principal axis frame:

$$\mathbf{I} = \begin{pmatrix} I_1 & 0 & 0 \\ 0 & I_2 & 0 \\ 0 & 0 & I_3 \end{pmatrix} \tag{6.13}$$

$$T_{\mathrm{rot}} = \frac{1}{2} (I_1 \omega_1^2 + I_2 \omega_2^2 + I_3 \omega_3^2) \tag{6.14}$$

Classification by symmetry:

- Spherical top: $I_1 = I_2 = I_3$
- Symmetric top: $I_1 = I_2 \neq I_3$ (prolate if $I_3 < I_1$, oblate if $I_3 > I_1$)
- Asymmetric top: $I_1 \neq I_2 \neq I_3$

6.1.5 Parallel Axis Theorem

If $\mathbf{I}^{\text{cm}}$ is the inertia tensor about the center of mass and $\mathbf{I}^O$ is about point O displaced by $\mathbf{a}$:

$$I_{ij}^O = I_{ij}^{\text{cm}} + M(a^2\delta_{ij} - a_i a_j) \tag{6.15}$$

This allows calculation of moments about any point once the center of mass values are known.

6.1.6 Angular Momentum and Kinetic Energy

The angular momentum about the center of mass is:

$$\mathbf{L} = \int \boldsymbol{\rho} \times \rho(\boldsymbol{\rho})\dot{\boldsymbol{\rho}} d^3r = \int \boldsymbol{\rho} \times \rho(\boldsymbol{\rho})(\boldsymbol{\omega} \times \boldsymbol{\rho}) d^3r = \mathbf{I} \cdot \boldsymbol{\omega} \tag{6.16}$$

In component form: $L_i = \sum_j I_{ij}\omega_j$

Note that $\mathbf{L}$ and $\boldsymbol{\omega}$ are generally not parallel unless $\boldsymbol{\omega}$ points along a principal axis.

The rotational kinetic energy can be expressed as:

$$T_{\text{rot}} = \frac{1}{2}\boldsymbol{\omega} \cdot \mathbf{L} \tag{6.17}$$

6.2 Euler Angles

6.2.1 Definition and Geometric Interpretation

Euler angles (ϕ, θ, ψ) parametrize the rotation from a space-fixed frame to a body-fixed frame through three successive rotations:

1. Rotate by ϕ about the z-axis: $(x, y, z) \to (x', y', z')$
2. Rotate by θ about the x'-axis: $(x', y', z') \to (x'', y'', z'')$
3. Rotate by ψ about the z''-axis: $(x'', y'', z'') \to (x_b, y_b, z_b)$

The complete rotation matrix is:

$$R = R_z(\psi)R_{x'}(\theta)R_z(\phi) \tag{6.18}$$

Explicitly:

$$R = \begin{pmatrix} \cos\psi\cos\phi - \cos\theta\sin\phi\sin\psi & \cos\psi\sin\phi + \cos\theta\cos\phi\sin\psi & \sin\psi\sin\theta \\ -\sin\psi\cos\phi - \cos\theta\sin\phi\cos\psi & -\sin\psi\sin\phi + \cos\theta\cos\phi\cos\psi & \cos\psi\sin\theta \\ \sin\theta\sin\phi & -\sin\theta\cos\phi & \cos\theta \end{pmatrix} \tag{6.19}$$

6.2.2 Angular Velocity in Terms of Euler Angles

The angular velocity vector has components in both space-fixed and body-fixed frames. The transformation between Euler angle rates and body-fixed components is:

$$\omega_1 = \dot{\phi} \sin\theta \sin\psi + \dot{\theta} \cos\psi \tag{6.20}$$

$$\omega_2 = \dot{\phi} \sin\theta \cos\psi - \dot{\theta} \sin\psi \tag{6.21}$$

$$\omega_3 = \dot{\phi} \cos\theta + \dot{\psi} \tag{6.22}$$

Or in matrix form:

$$\begin{pmatrix} \omega_1 \\ \omega_2 \\ \omega_3 \end{pmatrix} = \begin{pmatrix} \sin\theta \sin\psi & \cos\psi & 0 \\ \sin\theta \cos\psi & -\sin\psi & 0 \\ \cos\theta & 0 & 1 \end{pmatrix} \begin{pmatrix} \dot{\phi} \\ \dot{\theta} \\ \dot{\psi} \end{pmatrix} \tag{6.23}$$

6.2.3 Kinetic Energy in Euler Angles

For a symmetric top ($I_1 = I_2 \equiv I$), the rotational kinetic energy becomes:

$$T_{\text{rot}} = \frac{1}{2} I(\omega_1^2 + \omega_2^2) + \frac{1}{2} I_3 \omega_3^2 \tag{6.24}$$

$$= \frac{1}{2} I(\dot{\phi}^2 \sin^2\theta + \dot{\theta}^2) + \frac{1}{2} I_3(\dot{\phi} \cos\theta + \dot{\psi})^2 \tag{6.25}$$

This expression reveals that:

- ϕ and ψ are cyclic coordinates (absent from T if V doesn't depend on them)
- Their conjugate momenta are conserved

6.2.4 Singularities and Gimbal Lock

Euler angles have singularities when $\theta = 0$ or π. At these points:

- The ϕ and ψ rotations become indistinguishable
- The transformation matrix becomes singular

This is the "gimbal lock" problem in engineering.

Alternative parametrizations (quaternions, Rodrigues parameters) avoid these singularities but lack the intuitive geometric interpretation of Euler angles.

6.2.5 Lagrangian for a Rigid Body

The complete Lagrangian for a rigid body in a potential V is:

$$L = \frac{1}{2}M|\dot{\mathbf{R}}|^2 + \frac{1}{2}I(\dot{\phi}^2 \sin^2\theta + \dot{\theta}^2) + \frac{1}{2}I_3(\dot{\phi}\cos\theta + \dot{\psi})^2 - V \tag{6.26}$$

If V depends only on the center of mass position (uniform field), rotational and translational motions decouple.

6.2.6 Conserved Quantities

For a free rigid body ($V = 0$):

$$p_\phi = \frac{\partial L}{\partial \dot{\phi}} = I\sin^2\theta\dot{\phi} + I_3\cos\theta(\dot{\phi}\cos\theta + \dot{\psi}) = L_z \quad \text{(space-fixed)} \tag{6.27}$$

$$p_\psi = \frac{\partial L}{\partial \dot{\psi}} = I_3(\dot{\phi}\cos\theta + \dot{\psi}) = L_3 \quad \text{(body-fixed)} \tag{6.28}$$

Both components of angular momentum along the space z-axis and body 3-axis are conserved.

6.3 Symmetric and Asymmetric Tops

6.3.1 The Symmetric Top

For a symmetric top ($I_1 = I_2 = I \neq I_3$), the Lagrangian with no external torques is:

$$L = \frac{1}{2}I(\dot{\phi}^2 \sin^2\theta + \dot{\theta}^2) + \frac{1}{2}I_3(\dot{\phi}\cos\theta + \dot{\psi})^2 \tag{6.29}$$

The equations of motion from Lagrange's equations yield:

$$\frac{d}{dt}(I\sin^2\theta\dot{\phi} + I_3\cos\theta(\dot{\phi}\cos\theta + \dot{\psi})) = 0 \tag{6.30}$$

$$\frac{d}{dt}(I_3(\dot{\phi}\cos\theta + \dot{\psi})) = 0 \tag{6.31}$$

$$I\ddot{\theta} - I\sin\theta\cos\theta\dot{\phi}^2 + I_3\sin\theta(\dot{\phi}\cos\theta + \dot{\psi})\dot{\phi} = 0 \tag{6.32}$$

6.3.2 Regular Precession

The first two equations give conserved quantities:

$L_z = a = \text{constant}$ (total angular momentum projection on space z-axis) (6.33)

$L_3 = b = \text{constant}$ (projection on body symmetry axis) (6.34)

For regular precession, we seek solutions with θ = constant. The third equation becomes:

$$\sin\theta[a\cos\theta - (I\cos\theta + (I_3 - I)\frac{b}{I_3})] = 0 \tag{6.35}$$

This yields either $\theta = 0, \pi$ (trivial) or:

$$\cos\theta = \frac{(a-b)I}{(I-I_3)L} \tag{6.36}$$

where $L = |\mathbf{L}| = \sqrt{a^2\sin^2\theta + b^2\cos^2\theta}$.

6.3.3 Stability Analysis

The effective potential for θ motion is:

$$V_{\text{eff}}(\theta) = \frac{(a - b\cos\theta)^2}{2I\sin^2\theta} + \frac{b^2\cos^2\theta}{2I} \tag{6.37}$$

Stable precession requires $d^2V_{\text{eff}}/d\theta^2 > 0$ at the equilibrium angle. This leads to stability conditions involving the ratio of spin to precession rates.

6.3.4 The Heavy Symmetric Top

Adding gravity with the center of mass at distance h from the fixed point:

$$V = Mgh\cos\theta \tag{6.38}$$

The Lagrangian becomes:

$$L = \frac{1}{2}I(\dot{\phi}^2\sin^2\theta + \dot{\theta}^2) + \frac{1}{2}I_3(\dot{\phi}\cos\theta + \dot{\psi})^2 - Mgh\cos\theta \tag{6.39}$$

Now only p_ϕ and p_ψ are conserved. The energy is:

$$E = \frac{1}{2}I\dot{\theta}^2 + V_{\text{eff}}(\theta) \tag{6.40}$$

where:

$$V_{\text{eff}}(\theta) = \frac{(p_\phi - p_\psi \cos\theta)^2}{2I\sin^2\theta} + \frac{p_\psi^2}{2I_3} + Mgh\cos\theta \tag{6.41}$$

6.3.5 Nutation and Precession

The motion consists of:

- Precession: $\dot{\phi}$ varies as θ oscillates
- Spin: $\dot{\psi}$ about the symmetry axis
- Nutation: θ oscillates between turning points

For small nutations about stable precession:

$$\theta(t) = \theta_0 + A\cos(\omega_n t) \tag{6.42}$$

where ω_n is the nutation frequency.

6.3.6 The Asymmetric Top

For $I_1 \neq I_2 \neq I_3$, no analytical solution exists in general. The motion is described by Euler's equations in the body frame:

$$I_1\dot{\omega}_1 - (I_2 - I_3)\omega_2\omega_3 = N_1 \tag{6.43}$$
$$I_2\dot{\omega}_2 - (I_3 - I_1)\omega_3\omega_1 = N_2 \tag{6.44}$$
$$I_3\dot{\omega}_3 - (I_1 - I_2)\omega_1\omega_2 = N_3 \tag{6.45}$$

where N_i are external torque components.

For free motion ($N_i = 0$), the system is integrable with:

$$E = \frac{1}{2}(I_1\omega_1^2 + I_2\omega_2^2 + I_3\omega_3^2) = \text{constant} \tag{6.46}$$
$$L^2 = I_1^2\omega_1^2 + I_2^2\omega_2^2 + I_3^2\omega_3^2 = \text{constant} \tag{6.47}$$

The solution involves Jacobi elliptic functions.

6.4 System of Particles with Constraints

6.4.1 From Particles to Rigid Bodies

A rigid body is the limit of a system of particles with distance constraints:

$$f_{ij} = |\mathbf{r}_i - \mathbf{r}_j|^2 - d_{ij}^2 = 0 \tag{6.48}$$

For N particles, there are $N(N-1)/2$ possible constraints, but only $3N-6$ are independent (the number of internal degrees of freedom).

6.4.2 Holonomic Constraints and Virtual Work

For holonomic constraints $f_\alpha(\mathbf{r}_1, \ldots, \mathbf{r}_N, t) = 0$, the principle of virtual work gives:

$$\sum_i (\mathbf{F}_i - m_i \ddot{\mathbf{r}}_i) \cdot \delta \mathbf{r}_i = 0 \tag{6.49}$$

subject to:

$$\sum_i \frac{\partial f_\alpha}{\partial \mathbf{r}_i} \cdot \delta \mathbf{r}_i = 0 \tag{6.50}$$

Using Lagrange multipliers λ_α:

$$m_i \ddot{\mathbf{r}}_i = \mathbf{F}_i + \sum_\alpha \lambda_\alpha \frac{\partial f_\alpha}{\partial \mathbf{r}_i} \tag{6.51}$$

The multipliers represent constraint forces.

6.4.3 Example: The Physical Pendulum

A rigid body pivoted at point O (not the center of mass) in a gravitational field:

- Constraint: One point fixed
- Degrees of freedom: 3 (rotation about fixed point)
- Use Euler angles with origin at pivot

The Lagrangian is:

$$L = \frac{1}{2} \boldsymbol{\omega}^T \mathbf{I}^O \boldsymbol{\omega} - Mgh\cos\theta \tag{6.52}$$

where h is the distance from pivot to center of mass and $\mathbf{I}^O$ is the inertia tensor about the pivot.

For small oscillations about stable equilibrium:

$$\omega^2 = \frac{Mgh}{I_\theta} \tag{6.53}$$

where I_θ is the moment of inertia about the horizontal axis through O.

6.4.4 Coupled Rigid Bodies

Consider two rigid bodies connected by constraints (hinges, gears, etc.). The configuration space is the product of individual configuration spaces modulo constraints.

Example: Double Pendulum with Rigid Bodies

- Two rigid bodies connected at a pivot
- $6 + 6 - 3 = 9$ degrees of freedom
- Lagrangian couples translational and rotational motion

6.4.5 Rolling Constraints

A sphere rolling without slipping on a surface has non-holonomic constraints:

$$\mathbf{v}_{\text{contact}} = R\boldsymbol{\omega} \tag{6.54}$$

These constraints are non-integrable—the final orientation depends on the path taken. The Lagrangian method requires careful treatment:

$$L = T - V - \boldsymbol{\lambda} \cdot (\mathbf{v} - R\boldsymbol{\omega}) \tag{6.55}$$

where $\boldsymbol{\lambda}$ are Lagrange multiplier functions.

6.4.6 Separation of Timescales

In many systems, rigid body constraints can be viewed as the limit of stiff springs:

$$V_{\text{constraint}} = \frac{k}{2} \sum (|\mathbf{r}_i - \mathbf{r}_j| - d_{ij})^2 \tag{6.56}$$

As $k \to \infty$:

- Fast vibrations about rigid configuration
- Slow rigid body motion

Separation allows treating rigidity as a constraint rather than a force.

Summary

Rigid body mechanics showcases the power of the Lagrangian formulation in handling complex constrained systems. The automatic incorporation of rigidity constraints through appropriate coordinates—Euler angles for orientation—dramatically simplifies the analysis compared to tracking individual particle motions.

Key insights from rigid body mechanics:

1. **Separation of Motion**: The natural division into translational and rotational motion reflects the underlying symmetry of rigid constraints.
2. **Inertia Tensor**: This fundamental quantity encodes how mass distribution affects rotational dynamics, analogous to mass in translational motion.
3. **Conservation Laws**: Angular momentum components in both space-fixed and body-fixed frames provide powerful constraints on motion.
4. **Rich Dynamics**: Even simple systems like the symmetric top exhibit complex behavior—precession, nutation, and spin—arising from the interplay of conservation laws.
5. **Geometric Structure**: The configuration space $SO(3) \times \mathbb{R}^3$ has non-trivial topology, leading to phenomena like gimbal lock and requiring careful parametrization.

The rigid body approximation bridges microscopic particle mechanics and macroscopic continuum mechanics. The techniques developed here—especially the treatment of constraints and the use of appropriate generalized coordinates—extend naturally to more complex systems including deformable bodies and fields, which we explore in subsequent chapters.

Chapter 7

Continuum Mechanics

The transition from discrete particles to continuous media represents one of the most profound applications of the principle of least action. In continuum mechanics, we replace sums over particles with integrals over space, and the Lagrangian becomes a Lagrangian density integrated over the system volume. This approach unifies the treatment of fluids, elastic solids, and waves, revealing deep connections between seemingly disparate phenomena.

The variational formulation of continuum mechanics offers unique advantages. It automatically ensures conservation laws through Noether's theorem, provides systematic approximation methods, and extends naturally to relativistic field theories. From ocean waves to seismic propagation, from blood flow to stellar dynamics, continuum mechanics governs phenomena across vast scales. This chapter develops the variational approach systematically, demonstrating how the principle of least action illuminates the physics of continuous media.

7.1 Variational Formulation for Fluids

7.1.1 From Discrete to Continuous Systems

Consider the limiting process from N particles to a continuous medium. For particles with positions $\mathbf{r}_i(t)$ and masses m_i, the Lagrangian is:

$$L = \sum_i \left[\frac{1}{2}m_i|\dot{\mathbf{r}}_i|^2 - V_i\right] \tag{7.1}$$

As $N \to \infty$ with particles filling a volume, we introduce:

$$\text{Mass density}: \quad \rho(\mathbf{r}, t) = \lim_{\Delta V \to 0} \frac{\sum_{i \in \Delta V} m_i}{\Delta V} \tag{7.2}$$

$$\text{Velocity field}: \quad \mathbf{v}(\mathbf{r}, t) = \text{velocity at position } \mathbf{r} \tag{7.3}$$

$$\text{Lagrangian density}: \quad \mathcal{L} = \lim_{\Delta V \to 0} \frac{L_{\Delta V}}{\Delta V} \tag{7.4}$$

The action becomes:

$$S = \int \int \mathcal{L}(\mathbf{r}, t)\, d^3r\, dt \tag{7.5}$$

7.1.2 Lagrangian and Eulerian Descriptions

Two frameworks exist for describing fluid motion:

Lagrangian Description: Follow individual fluid parcels

- Label parcels by initial position $\mathbf{a}$
- Trajectory: $\mathbf{r} = \boldsymbol{\chi}(\mathbf{a}, t)$
- Velocity: $\mathbf{v} = \frac{\partial \boldsymbol{\chi}}{\partial t}\big|_{\mathbf{a}}$

Eulerian Description: Focus on fixed points in space

- Velocity field: $\mathbf{v}(\mathbf{r}, t)$
- Density field: $\rho(\mathbf{r}, t)$
- Pressure field: $p(\mathbf{r}, t)$

The material derivative connects these descriptions:

$$\frac{D}{Dt} = \frac{\partial}{\partial t} + \mathbf{v} \cdot \nabla \tag{7.6}$$

7.1.3 Action Principle for Ideal Fluids

For an ideal (inviscid, non-heat-conducting) fluid, Lin introduced the action:

$$S = \int \int \rho[\varepsilon(\rho, s) + \frac{1}{2}|\mathbf{v}|^2 - \Phi]\, d^3r\, dt \tag{7.7}$$

where:

- $\varepsilon(\rho, s)$ = internal energy per unit mass
- s = entropy per unit mass (conserved along streamlines)
- Φ = external potential (e.g., gravitational)

This action must be varied subject to constraints:

$$\text{Mass conservation}: \quad \frac{\partial \rho}{\partial t} + \nabla \cdot (\rho \mathbf{v}) = 0 \tag{7.8}$$

$$\text{Entropy conservation}: \quad \frac{Ds}{Dt} = 0 \tag{7.9}$$

7.1.4 Clebsch Representation

To handle the constraints, Clebsch introduced potentials ϕ, α, β:

$$\mathbf{v} = \nabla\phi + \alpha\nabla\beta \tag{7.10}$$

This automatically satisfies the vorticity equation. The action becomes:

$$S = \int\int \left\{ \rho \left[\frac{1}{2}|\nabla\phi + \alpha\nabla\beta|^2 - \varepsilon(\rho, s) - \Phi \right] - \phi \left[\frac{\partial \rho}{\partial t} + \nabla \cdot (\rho \mathbf{v}) \right] - \gamma\rho\frac{Ds}{Dt} \right\} d^3r\, dt \tag{7.11}$$

where ϕ and γ are Lagrange multipliers.

7.1.5 Derivation of Euler Equations

Varying the action with respect to $\phi, \alpha, \beta, \rho$, and s yields:

$$\delta S/\delta\phi: \quad \frac{\partial \rho}{\partial t} + \nabla \cdot (\rho \mathbf{v}) = 0 \quad \text{(continuity)} \tag{7.12}$$

$$\delta S/\delta\alpha: \quad \frac{\partial (\rho\beta)}{\partial t} + \nabla \cdot (\rho\beta \mathbf{v}) = 0 \tag{7.13}$$

$$\delta S/\delta\beta: \quad \frac{\partial (\rho\alpha)}{\partial t} + \nabla \cdot (\rho\alpha \mathbf{v}) + \rho = 0 \tag{7.14}$$

$$\delta S/\delta\rho: \quad \frac{1}{2}|\mathbf{v}|^2 - \varepsilon - \rho\left.\frac{\partial \varepsilon}{\partial \rho}\right|_s - \Phi + \frac{\partial \phi}{\partial t} + \mathbf{v}\cdot\nabla\phi + \gamma\frac{Ds}{Dt} = 0 \tag{7.15}$$

$$\delta S/\delta s: \quad \rho\left[\left.\frac{\partial \varepsilon}{\partial s}\right|_\rho + \frac{D\gamma}{Dt} \right] = 0 \tag{7.16}$$

7.1.6 The Euler Equations

Combining these variations and using thermodynamic relations:

$$d\varepsilon = Tds - pd(1/\rho) \tag{7.17}$$

$$\text{Enthalpy}: \quad h = \varepsilon + p/\rho \tag{7.18}$$

We obtain Euler's equations:

$$\frac{\partial \mathbf{v}}{\partial t} + (\mathbf{v} \cdot \nabla)\mathbf{v} = -\frac{1}{\rho}\nabla p - \nabla\Phi \tag{7.19}$$

With the continuity equation:

$$\frac{\partial \rho}{\partial t} + \nabla \cdot (\rho \mathbf{v}) = 0 \tag{7.20}$$

And for adiabatic flow:

$$\frac{Ds}{Dt} = 0 \tag{7.21}$$

These form a complete system for ideal fluid dynamics.

7.2 Euler and Navier-Stokes Equations

7.2.1 Conservation Form of Euler Equations

The Euler equations can be written in conservation form:

$$\frac{\partial \rho}{\partial t} + \nabla \cdot (\rho \mathbf{v}) = 0 \quad \text{(mass)} \tag{7.22}$$

$$\frac{\partial (\rho \mathbf{v})}{\partial t} + \nabla \cdot (\rho \mathbf{v} \otimes \mathbf{v} + p\mathbf{I}) = \rho \mathbf{g} \quad \text{(momentum)} \tag{7.23}$$

$$\frac{\partial (\rho E)}{\partial t} + \nabla \cdot [(\rho E + p)\mathbf{v}] = \rho \mathbf{g} \cdot \mathbf{v} \quad \text{(energy)} \tag{7.24}$$

where $E = \varepsilon + \frac{1}{2}|\mathbf{v}|^2$ is total energy per unit mass.

7.2.2 Vorticity Dynamics

Taking the curl of Euler's equation yields the vorticity equation:

$$\frac{D\boldsymbol{\omega}}{Dt} = (\boldsymbol{\omega} \cdot \nabla)\mathbf{v} + \boldsymbol{\omega}(\nabla \cdot \mathbf{v}) - \frac{1}{\rho^2}\nabla\rho \times \nabla p \tag{7.25}$$

where $\boldsymbol{\omega} = \nabla \times \mathbf{v}$ is vorticity.

For barotropic flow ($p = p(\rho)$):

$$\frac{D\boldsymbol{\omega}}{Dt} = (\boldsymbol{\omega} \cdot \nabla)\mathbf{v} \tag{7.26}$$

This shows vortex lines move with the fluid and can be stretched but not created.

7.2.3 Kelvin's Circulation Theorem

The circulation around a material contour $C(t)$ is:

$$\Gamma = \oint_{C(t)} \mathbf{v} \cdot d\mathbf{l} \tag{7.27}$$

Kelvin's theorem states:

$$\frac{D\Gamma}{Dt} = -\oint_{C(t)} \frac{1}{\rho}\nabla p \cdot d\mathbf{l} \tag{7.28}$$

For barotropic flow, circulation is conserved along material contours.

7.2.4 Including Viscosity: Navier-Stokes Equations

Real fluids exhibit viscosity—internal friction that dissipates energy. The stress tensor becomes:

$$\sigma_{ij} = -p\delta_{ij} + \tau_{ij} \tag{7.29}$$

where the viscous stress tensor is:

$$\tau_{ij} = \mu\left(\frac{\partial v_i}{\partial x_j} + \frac{\partial v_j}{\partial x_i}\right) + \lambda\delta_{ij}\nabla \cdot \mathbf{v} \tag{7.30}$$

For incompressible flow ($\nabla \cdot \mathbf{v} = 0$), the Navier-Stokes equations are:

$$\rho\left[\frac{\partial \mathbf{v}}{\partial t} + (\mathbf{v} \cdot \nabla)\mathbf{v}\right] = -\nabla p + \mu\nabla^2\mathbf{v} + \rho\mathbf{g} \tag{7.31}$$

$$\nabla \cdot \mathbf{v} = 0 \tag{7.32}$$

7.2.5 Variational Principle for Viscous Flow

Viscous effects don't derive from a standard Lagrangian. However, Onsager's variational principle for irreversible processes yields:

$$\delta \int [L - \Phi] dt = 0 \tag{7.33}$$

where L is the Lagrangian for ideal flow and Φ is the dissipation function:

$$\Phi = \frac{1}{2} \int \tau_{ij} \left(\frac{\partial v_i}{\partial x_j} + \frac{\partial v_j}{\partial x_i} \right) d^3 r \tag{7.34}$$

This "Lagrangian with friction" approach extends to other dissipative systems.

7.2.6 Boundary Conditions and Well-Posedness

The Euler and Navier-Stokes equations require boundary conditions:

Euler (inviscid):

- $\mathbf{v} \cdot \mathbf{n} = 0$ (no penetration at solid boundaries)
- Pressure specified at free surfaces

Navier-Stokes (viscous):

- $\mathbf{v} = 0$ (no-slip at solid boundaries)
- Stress continuity at interfaces

The mathematical character changes:

- Euler: Hyperbolic (wave-like)
- Navier-Stokes: Parabolic (diffusive)

7.3 Wave Propagation in Continuous Media

7.3.1 Small Amplitude Waves

Consider small perturbations about equilibrium:

$$\rho = \rho_0 + \rho' \tag{7.35}$$

$$\mathbf{v} = \mathbf{v}' \tag{7.36}$$

$$p = p_0 + p' \tag{7.37}$$

Linearizing Euler's equations:

$$\frac{\partial \rho'}{\partial t} + \rho_0 \nabla \cdot \mathbf{v}' = 0 \tag{7.38}$$

$$\rho_0 \frac{\partial \mathbf{v}'}{\partial t} + \nabla p' = 0 \tag{7.39}$$

With the equation of state $p' = c^2 \rho'$ (c = sound speed), we obtain:

$$\frac{\partial^2 \rho'}{\partial t^2} = c^2 \nabla^2 \rho' \tag{7.40}$$

This is the wave equation with propagation speed $c = \sqrt{\left.\frac{\partial p}{\partial \rho}\right|_s}$.

7.3.2 Lagrangian Density for Waves

The wave equation derives from the Lagrangian density:

$$\mathcal{L} = \frac{1}{2} \left[\rho_0 \left(\frac{\partial \phi}{\partial t} \right)^2 - \rho_0 c^2 |\nabla \phi|^2 \right] \tag{7.41}$$

where ϕ is the velocity potential ($\mathbf{v} = \nabla \phi$).

The action is:

$$S = \int \int \mathcal{L} \, d^3 r \, dt \tag{7.42}$$

Varying with respect to ϕ yields:

$$\frac{\partial^2 \phi}{\partial t^2} = c^2 \nabla^2 \phi \tag{7.43}$$

7.3.3 Energy and Momentum of Waves

Noether's theorem applied to the wave Lagrangian gives:

$$\text{Energy density}: \quad \varepsilon = \frac{1}{2}\rho_0\left[\left(\frac{\partial\phi}{\partial t}\right)^2 + c^2|\nabla\phi|^2\right] \tag{7.44}$$

$$\text{Energy flux (Poynting vector)}: \quad \mathbf{S} = -\rho_0\frac{\partial\phi}{\partial t}\nabla\phi \tag{7.45}$$

$$\text{Momentum density}: \quad \mathbf{g} = \rho_0\frac{\partial\phi}{\partial t}\nabla\phi = -\frac{\mathbf{S}}{c^2} \tag{7.46}$$

Conservation laws:

$$\frac{\partial\varepsilon}{\partial t} + \nabla\cdot\mathbf{S} = 0 \tag{7.47}$$

$$\frac{\partial\mathbf{g}}{\partial t} + \nabla\cdot\mathbf{T} = 0 \tag{7.48}$$

where T_{ij} is the stress tensor.

7.3.4 Dispersion and Group Velocity

For waves in dispersive media, the relation $\omega = \omega(\mathbf{k})$ is nonlinear. The Lagrangian density becomes:

$$\mathcal{L} = \frac{1}{2}\rho_0\left(\frac{\partial\phi}{\partial t}\right)^2 - \frac{1}{2}\rho_0\phi\hat{\mathcal{D}}[\phi] \tag{7.49}$$

where $\hat{\mathcal{D}}$ is a pseudo-differential operator.

The group velocity is:

$$\mathbf{v}_g = \frac{\partial\omega}{\partial\mathbf{k}} \tag{7.50}$$

Energy propagates at the group velocity, while phase propagates at $\mathbf{v}_p = \omega/k\hat{\mathbf{k}}$.

7.3.5 Nonlinear Waves

For finite amplitude waves, nonlinear terms become important:

$$\frac{\partial u}{\partial t} + u\frac{\partial u}{\partial x} = c^2\frac{\partial u}{\partial x} \tag{7.51}$$

This leads to wave steepening and shock formation. The variational formulation helps identify conserved quantities even when shocks develop.

7.3.6 Surface Waves

Water waves involve a free surface with boundary conditions:

$$\text{Kinematic}: \quad \frac{D\eta}{Dt} = w \quad \text{at } z = \eta(x, y, t) \tag{7.52}$$

$$\text{Dynamic}: \quad p = p_{\text{atm}} \quad \text{at } z = \eta \tag{7.53}$$

The Lagrangian includes surface terms:

$$L = \int\int\int \frac{1}{2}\rho|\nabla\phi|^2 \, dV - \frac{1}{2}\rho g \int\int \eta^2 \, dA \tag{7.54}$$

Leading to various wave types:

- Deep water: $\omega^2 = gk$
- Shallow water: $\omega^2 = ghk^2$
- Capillary: $\omega^2 = (\sigma/\rho)k^3$

7.4 Elasticity and Principle of Least Action

7.4.1 Kinematics of Deformation

An elastic body deforms under applied forces. The deformation is described by:

$$\mathbf{x} = \boldsymbol{\chi}(\mathbf{X}, t) \tag{7.55}$$

where $\mathbf{X}$ labels material points (Lagrangian coordinates) and $\mathbf{x}$ is current position.

The deformation gradient is:

$$F_{ij} = \frac{\partial x_i}{\partial X_j} \tag{7.56}$$

Key kinematic quantities:

$$\text{Displacement}: \quad \mathbf{u} = \mathbf{x} - \mathbf{X} \tag{7.57}$$

$$\text{Strain tensor}: \quad E_{ij} = \frac{1}{2}(F_{ki}F_{kj} - \delta_{ij}) \tag{7.58}$$

$$\text{Small strain}: \quad \varepsilon_{ij} = \frac{1}{2}\left(\frac{\partial u_i}{\partial X_j} + \frac{\partial u_j}{\partial X_i}\right) \tag{7.59}$$

7.4.2 Elastic Energy and Constitutive Relations

The stored elastic energy per unit reference volume is:

$$W = W(E_{ij}) \tag{7.60}$$

For small strains and isotropic materials:

$$W = \frac{\lambda}{2}(\varepsilon_{kk})^2 + \mu\varepsilon_{ij}\varepsilon_{ij} \tag{7.61}$$

where λ and μ are Lamé parameters.

The stress-strain relation (Hooke's law):

$$\sigma_{ij} = \frac{\partial W}{\partial \varepsilon_{ij}} = \lambda\varepsilon_{kk}\delta_{ij} + 2\mu\varepsilon_{ij} \tag{7.62}$$

7.4.3 Lagrangian for Elastodynamics

The Lagrangian density in material coordinates:

$$\mathcal{L} = \frac{1}{2}\rho_0\left|\frac{\partial \mathbf{u}}{\partial t}\right|^2 - W(\varepsilon_{ij}) - \rho_0\Phi \tag{7.63}$$

where ρ_0 is reference density.

The action:

$$S = \int\int \mathcal{L}\, d^3X\, dt \tag{7.64}$$

7.4.4 Equations of Motion

Varying the action with respect to u_i:

$$\rho_0\frac{\partial^2 u_i}{\partial t^2} = \frac{\partial}{\partial X_j}\left(\frac{\partial W}{\partial \varepsilon_{ij}}\right) + \rho_0 f_i \tag{7.65}$$

For linear elasticity:

$$\rho_0 \frac{\partial^2 u_i}{\partial t^2} = \frac{\partial \sigma_{ij}}{\partial X_j} + \rho_0 f_i \tag{7.66}$$

In terms of displacements:

$$\rho_0 \frac{\partial^2 \mathbf{u}}{\partial t^2} = (\lambda + \mu)\nabla(\nabla \cdot \mathbf{u}) + \mu \nabla^2 \mathbf{u} + \rho_0 \mathbf{f} \tag{7.67}$$

7.4.5 Elastic Waves

The equations support two wave types:

Longitudinal (P-waves):

- Displacement parallel to propagation
- Speed: $c_p = \sqrt{(\lambda + 2\mu)/\rho_0}$
- $\nabla \times \mathbf{u} = 0$

Transverse (S-waves):

- Displacement perpendicular to propagation
- Speed: $c_s = \sqrt{\mu/\rho_0}$
- $\nabla \cdot \mathbf{u} = 0$

Generally $c_p > c_s$, so P-waves arrive first in seismology.

7.4.6 Variational Principles in Elastostatics

For static equilibrium, the principle of minimum potential energy states:

$$\delta \Pi = \delta \int [W - \rho_0 \Phi - \mathbf{t} \cdot \mathbf{u}] dV = 0 \tag{7.68}$$

where $\mathbf{t}$ are surface tractions.

This yields:

- Equilibrium equations in the interior
- Natural boundary conditions on surfaces

Alternative principles:

- Hellinger-Reissner: Mixed formulation with stress and displacement
- Hu-Washizu: Three-field with stress, strain, and displacement

7.4.7 Finite Deformations

For large deformations, the theory becomes geometrically nonlinear:

$$\text{Green-Lagrange strain}: \quad E_{ij} = \frac{1}{2}(F_{ki}F_{kj} - \delta_{ij}) \tag{7.69}$$

$$\text{Second Piola-Kirchhoff stress}: \quad S_{ij} = \frac{\partial W}{\partial E_{ij}} \tag{7.70}$$

The equilibrium equation in material form:

$$\frac{\partial}{\partial X_j}(F_{ik}S_{kj}) + \rho_0 b_i = 0 \tag{7.71}$$

Constitutive models for finite deformations:

- Neo-Hookean: $W = \frac{\mu}{2}(I_1 - 3) + \frac{\lambda}{2}(J-1)^2$
- Mooney-Rivlin: $W = C_1(I_1 - 3) + C_2(I_2 - 3)$

where I_1, I_2 are strain invariants and $J = \det(F)$.

Summary

Continuum mechanics demonstrates the power and versatility of variational principles in handling systems with infinite degrees of freedom. The transition from discrete particles to continuous fields transforms the Lagrangian into a Lagrangian density, but the fundamental variational structure remains.

Key insights from this chapter:

1. **Unified Framework**: The principle of least action provides a common foundation for fluids, elastic solids, and wave phenomena, revealing deep connections between seemingly different systems.

2. **Conservation Laws**: Noether's theorem automatically yields conservation of mass, momentum, energy, and vorticity from symmetries of the action.

3. **Systematic Approximations**: The variational formulation facilitates perturbation methods, from linearization for small amplitude waves to multiple-scale analysis for nonlinear phenomena.

4. **Boundary Conditions**: Natural boundary conditions emerge from the variational principle, ensuring mathematical well-posedness and physical realizability.

5. **Extension to Dissipative Systems**: While ideal systems have standard Lagrangians, dissipative effects can be incorporated through Onsager's principle or Lagrangians with friction.

The continuum limit also prepares us for field theory. The mathematical structures developed here—Lagrangian densities, field variations, conservation laws from continuous symmetries—carry over directly to electromagnetic fields, quantum fields, and gravitational fields. The next chapter explores classical field theory, where these ideas reach their full expression in describing the fundamental forces of nature.

Chapter 8

Classical Field Theory

Classical field theory represents the culmination of classical physics and the gateway to modern physics. Fields—quantities defined at every point in spacetime—provide the natural description for electromagnetic phenomena, gravitational effects, and ultimately all fundamental interactions. The principle of least action, when applied to fields, yields a powerful framework that unifies diverse phenomena and naturally incorporates special relativity.

This chapter develops field theory from first principles, showing how the variational approach transforms our understanding of electromagnetism. We'll see how Maxwell's equations, originally discovered through painstaking experimental work, emerge naturally from a simple action principle. The covariant formulation not only elegantly incorporates special relativity but also provides the template for gauge theories in particle physics and general relativity. From radio waves to light, from electrical circuits to relativistic particle beams, field theory governs phenomena across vast scales of space and time.

8.1 Scalar and Vector Fields

8.1.1 From Particles to Fields

The transition from particle mechanics to field theory parallels our earlier passage from discrete to continuous systems, but with a crucial difference. While a continuous medium like a fluid still consists of particles (molecules), fields are truly fundamental entities existing at every spacetime point.

Consider the progression:

- Single particle: $q(t)$ — 1 degree of freedom
- N particles: $q_i(t)$ — N degrees of freedom
- Continuous medium: $\rho(\mathbf{r}, t), \mathbf{v}(\mathbf{r}, t)$ — ∞ degrees of freedom
- Field: $\phi(\mathbf{r}, t)$ or $\mathbf{A}(\mathbf{r}, t)$ — ∞ degrees of freedom

The key distinction: fields aren't merely convenient descriptions of many-particle systems but fundamental physical entities in their own right.

8.1.2 Scalar Fields

A scalar field $\phi(\mathbf{r}, t)$ assigns a number to each spacetime point. Physical examples include:

- Temperature $T(\mathbf{r}, t)$
- Pressure $p(\mathbf{r}, t)$
- Gravitational potential $\Phi(\mathbf{r}, t)$
- Quantum wave function $\psi(\mathbf{r}, t)$ (complex scalar)

The dynamics of a scalar field follow from a Lagrangian density:

$$\mathcal{L} = \mathcal{L}(\phi, \partial_\mu \phi, x^\mu) \tag{8.1}$$

where $\partial_\mu = \partial / \partial x^\mu$ and $x^\mu = (ct, x, y, z)$.

8.1.3 The Klein-Gordon Field

The simplest relativistic scalar field has Lagrangian density:

$$\mathcal{L} = \frac{1}{2}\partial_\mu \phi \partial^\mu \phi - \frac{1}{2}m^2\phi^2 \tag{8.2}$$

Using the metric signature $(+, -, -, -)$:

$$\partial_\mu \phi \partial^\mu \phi = \frac{1}{c^2}\left(\frac{\partial \phi}{\partial t}\right)^2 - |\nabla \phi|^2 \tag{8.3}$$

The action is:

$$S = \int \mathcal{L}\, d^4x = \int\int\int\int \mathcal{L}\, dx\, dy\, dz\, c\, dt \tag{8.4}$$

8.1.4 Field Equations from Variational Principle

Varying the action with respect to $\phi(\mathbf{r}, t)$:

$$\delta S = \int \left[\frac{\partial \mathcal{L}}{\partial \phi} \delta\phi + \frac{\partial \mathcal{L}}{\partial(\partial_\mu \phi)} \delta(\partial_\mu \phi) \right] d^4x \tag{8.5}$$

Integrating by parts:

$$\delta S = \int \left[\frac{\partial \mathcal{L}}{\partial \phi} - \partial_\mu \left(\frac{\partial \mathcal{L}}{\partial(\partial_\mu \phi)} \right) \right] \delta\phi \, d^4x \tag{8.6}$$

For stationary action:

$$\frac{\partial \mathcal{L}}{\partial \phi} - \partial_\mu \left(\frac{\partial \mathcal{L}}{\partial(\partial_\mu \phi)} \right) = 0 \tag{8.7}$$

This is the Euler-Lagrange equation for fields.

8.1.5 Klein-Gordon Equation

For the Klein-Gordon Lagrangian:

$$\frac{\partial \mathcal{L}}{\partial \phi} = -m^2 \phi \tag{8.8}$$

$$\frac{\partial \mathcal{L}}{\partial(\partial_\mu \phi)} = \partial^\mu \phi \tag{8.9}$$

The field equation becomes:

$$\Box \phi + m^2 \phi = 0 \tag{8.10}$$

where $\Box = \partial_\mu \partial^\mu = \frac{1}{c^2}\frac{\partial^2}{\partial t^2} - \nabla^2$ is the d'Alembertian operator.

For $m = 0$, this reduces to the wave equation with speed c.

8.1.6 Vector Fields

A vector field $A(\mathbf{r}, t)$ assigns a vector to each spacetime point. In component notation: $A^\mu(x)$ where $\mu = 0, 1, 2, 3$.

Physical examples:

- Electromagnetic four-potential $A^\mu = (\phi/c, \mathbf{A})$

- Velocity field in fluid dynamics
- Displacement field in elasticity

The Lagrangian density for vector fields typically involves:

$$\mathcal{L} = \mathcal{L}(A^\mu, \partial_\nu A^\mu, x) \tag{8.11}$$

8.1.7 Gauge Invariance

Vector fields often possess gauge symmetry—transformations that leave physics unchanged:

$$A^\mu \to A^\mu + \partial^\mu \chi \tag{8.12}$$

where χ is an arbitrary scalar function.

This redundancy in description reflects that not all components of A^μ are physical. The Lagrangian must be constructed to respect gauge invariance, leading to profound consequences for the structure of physical theories.

8.2 Electromagnetism and Maxwell Action

8.2.1 The Electromagnetic Field Tensor

From the four-potential $A^\mu = (\phi/c, \mathbf{A})$, we construct the antisymmetric field tensor:

$$F^{\mu\nu} = \partial^\mu A^\nu - \partial^\nu A^\mu \tag{8.13}$$

In matrix form:

$$F^{\mu\nu} = \begin{pmatrix} 0 & -E_x/c & -E_y/c & -E_z/c \\ E_x/c & 0 & B_z & -B_y \\ E_y/c & -B_z & 0 & B_x \\ E_z/c & B_y & -B_x & 0 \end{pmatrix} \tag{8.14}$$

The electric and magnetic fields are:

$$E_i = cF^{0i} = -\frac{\partial A_i}{\partial t} - \partial_i \phi \tag{8.15}$$

$$B_i = \frac{1}{2}\varepsilon_{ijk} F^{jk} = (\nabla \times \mathbf{A})_i \tag{8.16}$$

8.2.2 Gauge Invariance of $F^{\mu\nu}$

Under gauge transformations $A^\mu \to A^\mu + \partial^\mu \chi$:

$$F^{\mu\nu} \to \partial^\mu (A^\nu + \partial^\nu \chi) - \partial^\nu (A^\mu + \partial^\mu \chi) \tag{8.17}$$
$$= F^{\mu\nu} + \partial^\mu \partial^\nu \chi - \partial^\nu \partial^\mu \chi \tag{8.18}$$
$$= F^{\mu\nu} \tag{8.19}$$

The field tensor is gauge invariant, ensuring that $\mathbf{E}$ and $\mathbf{B}$ are physical observables.

8.2.3 The Maxwell Action

The electromagnetic field action is:

$$S_{\text{field}} = -\frac{1}{4\mu_0 c} \int F_{\mu\nu} F^{\mu\nu} \, d^4x \tag{8.20}$$

Expanding:

$$F_{\mu\nu} F^{\mu\nu} = 2\left(B^2 - \frac{E^2}{c^2}\right) \tag{8.21}$$

So:

$$S_{\text{field}} = \frac{1}{2\mu_0} \int\int\int\int \left(\frac{E^2}{c^2} - B^2\right) d^3r \, c \, dt \tag{8.22}$$

This is manifestly Lorentz invariant.

8.2.4 Interaction with Charges

A charged particle couples to the electromagnetic field through:

$$S_{\text{int}} = -\int j^\mu A_\mu \, d^4x \tag{8.23}$$

where $j^\mu = (c\rho, \mathbf{j})$ is the four-current density.

For a point charge:

$$j^\mu(x) = q \int u^\mu(\tau) \delta^4(x - x(\tau)) \, d\tau \tag{8.24}$$

where $u^\mu = dx^\mu / d\tau$ is the four-velocity.

8.2.5 Complete Electromagnetic Action

The total action combines field, matter, and interaction terms:

$$S = S_{\text{matter}} + S_{\text{field}} + S_{\text{int}} \tag{8.25}$$

For charged particles in an electromagnetic field:

$$S = \sum_a \int [-m_a c\sqrt{1 - v_a^2/c^2} - q_a(\phi - \mathbf{v}_a \cdot \mathbf{A})]dt + \frac{1}{2\mu_0} \int \int \int \int \left(\frac{E^2}{c^2} - B^2\right) d^3r\, c\, d \tag{8.26}$$

8.2.6 Gauge Fixing

The gauge freedom $A^\mu \to A^\mu + \partial^\mu \chi$ means the action has a continuous symmetry. To obtain unique solutions, we must fix the gauge:

Coulomb gauge: $\nabla \cdot \mathbf{A} = 0$

- Instantaneous Coulomb interaction
- Non-covariant

Lorenz gauge: $\partial_\mu A^\mu = 0$

- Manifestly covariant
- Symmetric treatment of space and time

Temporal gauge: $A^0 = 0$

- Used in Hamiltonian formulation
- Breaks manifest covariance

8.3 Covariant Formulation

8.3.1 Minkowski Spacetime

Special relativity unifies space and time into spacetime with metric:

$$ds^2 = c^2dt^2 - dx^2 - dy^2 - dz^2 = \eta_{\mu\nu}dx^\mu dx^\nu \tag{8.27}$$

where $\eta_{\mu\nu} = \text{diag}(1, -1, -1, -1)$.

Covariant (lower) and contravariant (upper) indices:

$$x_\mu = \eta_{\mu\nu} x^\nu \tag{8.28}$$

$$A_\mu = \eta_{\mu\nu} A^\nu \tag{8.29}$$

$$\partial^\mu = \eta^{\mu\nu} \partial_\nu \tag{8.30}$$

8.3.2 Lorentz Transformations

Lorentz transformations preserve the metric:

$$x'^\mu = \Lambda^\mu{}_\nu x^\nu \tag{8.31}$$

with $\Lambda^\mu{}_\rho \eta_{\mu\nu} \Lambda^\nu{}_\sigma = \eta_{\rho\sigma}$.

Fields transform as:

- Scalar: $\phi'(x') = \phi(x)$
- Vector: $A'^\mu(x') = \Lambda^\mu{}_\nu A^\nu(x)$
- Tensor: $F'^{\mu\nu}(x') = \Lambda^\mu{}_\rho \Lambda^\nu{}_\sigma F^{\rho\sigma}(x)$

8.3.3 Covariant Derivatives and Equations

The field equations must be covariant—maintaining their form under Lorentz transformations.

Covariant Maxwell equations:

$$\partial_\mu F^{\mu\nu} = \mu_0 j^\nu \tag{8.32}$$

Bianchi identity:

$$\partial_\mu F^{*\mu\nu} = 0 \tag{8.33}$$

where $F^{*\mu\nu} = \frac{1}{2}\varepsilon^{\mu\nu\rho\sigma} F_{\rho\sigma}$ is the dual tensor.

8.3.4 Energy-Momentum Tensor

Noether's theorem applied to spacetime translation symmetry yields the energy-momentum tensor:

$$T^{\mu\nu} = \frac{\partial \mathcal{L}}{\partial(\partial_\mu \phi)} \partial^\nu \phi - \eta^{\mu\nu} \mathcal{L} \tag{8.34}$$

For the electromagnetic field:

$$T^{\mu\nu} = \frac{1}{\mu_0}\left[F^{\mu\rho}F^{\nu}{}_{\rho} + \frac{1}{4}\eta^{\mu\nu}F_{\rho\sigma}F^{\rho\sigma}\right] \tag{8.35}$$

Components have physical meaning:

- T^{00} = energy density
- T^{0i} = energy flux/c = momentum density $\times c$
- T^{ij} = momentum flux = stress tensor

8.3.5 Conservation Laws

The energy-momentum tensor satisfies:

$$\partial_\mu T^{\mu\nu} = -f^\nu \tag{8.36}$$

where f^ν is the force density. For free fields:

$$\partial_\mu T^{\mu\nu} = 0 \tag{8.37}$$

This yields:

- $\nu = 0$: Energy conservation
- $\nu = i$: Momentum conservation

8.3.6 Angular Momentum Tensor

The angular momentum tensor:

$$M^{\mu\nu\rho} = x^\nu T^{\mu\rho} - x^\rho T^{\mu\nu} + S^{\mu\nu\rho} \tag{8.38}$$

where $S^{\mu\nu\rho}$ is the spin tensor. For electromagnetic fields:

$$S^{\mu\nu\rho} = A^\nu F^{\mu\rho} - A^\rho F^{\mu\nu} \tag{8.39}$$

Conservation of angular momentum:

$$\partial_\mu M^{\mu\nu\rho} = 0 \tag{8.40}$$

8.4 Maxwell Equations as Euler-Lagrange Equations

8.4.1 Variation of the Maxwell Action

Starting with:

$$S = -\frac{1}{4\mu_0 c}\int F_{\mu\nu}F^{\mu\nu}\,d^4x - \int j^\mu A_\mu\,d^4x \tag{8.41}$$

We vary with respect to A_μ:

$$\delta S = -\frac{1}{2\mu_0 c}\int F^{\mu\nu}\delta F_{\mu\nu}\,d^4x - \int j^\mu \delta A_\mu\,d^4x \tag{8.42}$$

Since $\delta F_{\mu\nu} = \partial_\mu(\delta A_\nu) - \partial_\nu(\delta A_\mu)$:

$$\delta S = -\frac{1}{\mu_0 c}\int F^{\mu\nu}\partial_\mu(\delta A_\nu)\,d^4x - \int j^\mu \delta A_\mu\,d^4x \tag{8.43}$$

8.4.2 Integration by Parts

Integrating by parts and dropping surface terms:

$$\delta S = \frac{1}{\mu_0 c}\int (\partial_\mu F^{\mu\nu})\delta A_\nu\,d^4x - \int j^\nu \delta A_\nu\,d^4x \tag{8.44}$$

$$= \int \left[\frac{1}{\mu_0 c}\partial_\mu F^{\mu\nu} - j^\nu\right]\delta A_\nu\,d^4x \tag{8.45}$$

8.4.3 The Inhomogeneous Maxwell Equations

For stationary action, $\delta S = 0$ for all δA_ν:

$$\partial_\mu F^{\mu\nu} = \mu_0 j^\nu \tag{8.46}$$

Expanding in components:

- $\nu = 0$: $\nabla \cdot \mathbf{E} = \rho/\varepsilon_0$ (Gauss's law)
- $\nu = i$: $\nabla \times \mathbf{B} - \frac{1}{c^2}\frac{\partial \mathbf{E}}{\partial t} = \mu_0 \mathbf{j}$ (Ampère-Maxwell law)

8.4.4 The Homogeneous Maxwell Equations

The Bianchi identity $\partial_{[\mu}F_{\nu\rho]} = 0$ gives:

$$\partial_\mu F_{\nu\rho} + \partial_\nu F_{\rho\mu} + \partial_\rho F_{\mu\nu} = 0 \tag{8.47}$$

This yields:

- $\nabla \cdot \mathbf{B} = 0$ (no magnetic monopoles)
- $\nabla \times \mathbf{E} + \frac{\partial \mathbf{B}}{\partial t} = 0$ (Faraday's law)

These equations are automatically satisfied by $F_{\mu\nu} = \partial_\mu A_\nu - \partial_\nu A_\mu$.

8.4.5 Wave Equations for Potentials

In the Lorenz gauge $\partial_\mu A^\mu = 0$:

$$\partial_\mu F^{\mu\nu} = \partial_\mu(\partial^\mu A^\nu - \partial^\nu A^\mu) \tag{8.48}$$
$$= \Box A^\nu - \partial^\nu(\partial_\mu A^\mu) \tag{8.49}$$
$$= \Box A^\nu \tag{8.50}$$

So Maxwell's equations become:

$$\Box A^\nu = \mu_0 j^\nu \tag{8.51}$$

These are four wave equations with sources.

8.4.6 Green's Functions and Retarded Solutions

The solution to $\Box G(x - x') = \delta^4(x - x')$ is:

$$G_{\text{ret}}(x - x') = \frac{1}{4\pi|\mathbf{x} - \mathbf{x}'|}\delta(t - t' - |\mathbf{x} - \mathbf{x}'|/c) \tag{8.52}$$

This gives the retarded potentials:

$$A^\mu(x) = \mu_0 \int G_{\text{ret}}(x - x')j^\mu(x')\,d^4x' = \frac{\mu_0}{4\pi}\int \frac{j^\mu(t - |\mathbf{x} - \mathbf{x}'|/c, \mathbf{x}')}{|\mathbf{x} - \mathbf{x}'|}\,d^3x' \tag{8.53}$$

8.4.7 Conservation of Charge

Taking ∂_ν of Maxwell's equation:

$$\partial_\nu \partial_\mu F^{\mu\nu} = \mu_0 \partial_\nu j^\nu \tag{8.54}$$

But $\partial_\nu \partial_\mu F^{\mu\nu} = 0$ (antisymmetry), so:

$$\partial_\nu j^\nu = 0 \tag{8.55}$$

In 3+1 form:

$$\frac{\partial \rho}{\partial t} + \nabla \cdot \mathbf{j} = 0 \tag{8.56}$$

This continuity equation expresses local charge conservation—a consequence of gauge invariance via Noether's theorem.

Summary

Classical field theory represents the triumph of the variational principle in classical physics. Starting from simple symmetry requirements—Lorentz invariance and gauge invariance—the principle of least action yields Maxwell's equations in their full glory. This achievement demonstrates several profound insights:

1. **Fields as Fundamental Entities**: Unlike particles or continuous media, fields exist at every spacetime point as primary physical objects, not mere mathematical conveniences.

2. **Unification through Geometry**: The covariant formulation reveals electromagnetism's geometric nature. Electric and magnetic fields unite in the field tensor $F^{\mu\nu}$, while space and time merge in Minkowski spacetime.

3. **Symmetry Dictates Dynamics**: Gauge invariance isn't merely a mathematical nicety but fundamentally determines electromagnetic interactions. This principle extends to all modern gauge theories.

4. **Conservation from Symmetry**: Noether's theorem automatically yields conservation of energy, momentum, angular momentum, and charge from spacetime and gauge symmetries.

5. **Template for Modern Physics**: The structure developed here—gauge invariant Lagrangian, covariant field equations, conserved currents—provides the blueprint for the Standard Model of particle physics.

The variational formulation also provides practical advantages:

- Systematic derivation of field equations
- Manifest covariance ensuring relativistic invariance
- Clear identification of physical degrees of freedom
- Natural incorporation of interactions
- Direct path to quantization

As we transition to Part III, we'll see how these classical field theory concepts extend to quantum mechanics, relativity, and quantum field theory. The mathematical structures—Lagrangian densities, gauge invariance, covariant formulation—remain, but acquire new physical interpretations in the quantum realm. The principle of least action, which began with Fermat's light rays, thus provides the foundation for our most fundamental theories of nature.

Part III

Applications in Modern Physics

Part III explores how the principle of least action extends beyond classical physics into the revolutionary theories of the 20th and 21st centuries. We begin with relativity, where the action principle not only survives the transition from Newtonian to Einsteinian physics but emerges strengthened and clarified. We then explore quantum mechanics, where the classical action reappears in Feynman's path integral formulation, and quantum field theory, where the synthesis of quantum mechanics and special relativity finds its natural expression in terms of field actions. Finally, we examine how gauge principles and the action formulation guide the construction of the Standard Model and beyond.

Throughout Part III, we'll see that the principle of least action is not merely preserved in modern physics—it becomes even more central. While forces and particles lose their fundamental status, the action remains, providing the bridge between classical intuition and quantum reality, between special relativity and gravity, between observed symmetries and dynamical laws.

Chapter 9

Relativity

Einstein's theories of special and general relativity revolutionized our understanding of space, time, and gravity. Remarkably, both theories find their most elegant formulation through the principle of least action. In special relativity, the action for a free particle takes a surprisingly simple form that automatically ensures Lorentz invariance. In general relativity, gravity itself emerges from an action principle—the Einstein-Hilbert action—transforming Newton's mysterious force into the geometry of spacetime.

This chapter develops relativistic physics from the variational perspective, showing how the principle of least action naturally accommodates Einstein's insights. We'll see how the requirement of reparametrization invariance leads to the relativistic action, how fields propagate in curved spacetime, and how Einstein's field equations emerge from varying the total action of spacetime and matter. The geodesic equation, describing motion in curved spacetime, appears as the Euler-Lagrange equation for a particle's worldline—gravity as geometry rather than force.

9.1 Principle of Least Action in Special Relativity

9.1.1 The Postulates and Their Implications

Special relativity rests on two postulates:

1. The laws of physics take the same form in all inertial frames

2. The speed of light c is constant in all inertial frames

These imply that space and time must transform together under boosts between frames. The invariant interval:

$$ds^2 = c^2dt^2 - dx^2 - dy^2 - dz^2 \tag{9.1}$$

replaces the separate invariants of Newtonian physics (spatial distance and time interval).

For the principle of least action to respect these postulates, the action itself must be a Lorentz scalar—invariant under transformations between inertial frames.

9.1.2 The Search for a Relativistic Action

In Newtonian mechanics, the action for a free particle is:

$$S_{\text{Newton}} = \int \left[\frac{1}{2}mv^2\right] dt \tag{9.2}$$

This cannot be correct relativistically because:

- It's not Lorentz invariant
- It allows $v > c$
- The integrand depends on the frame through v

We need an action that:

- Is Lorentz invariant
- Reduces to the Newtonian form for $v \ll c$
- Prevents superluminal motion

9.1.3 The Relativistic Action

The key insight is to use the invariant interval ds as the basis for integration. For a massive particle, the action is:

$$S = -mc \int ds = -mc^2 \int \sqrt{1 - v^2/c^2}\, dt \tag{9.3}$$

This can be written in manifestly covariant form:

$$S = -mc \int ds = -mc \int \sqrt{dx^\mu dx_\mu} \tag{9.4}$$

where the integral is along the worldline.

9.1.4 Reparametrization Invariance

The action can be parametrized by any monotonic parameter λ:

$$S = -mc \int \sqrt{\dot{x}^\mu \dot{x}_\mu}\, d\lambda \tag{9.5}$$

where $\dot{x}^\mu = dx^\mu/d\lambda$. This reparametrization invariance is a gauge symmetry of the worldline.

An equivalent action that makes calculations easier:

$$S = \frac{1}{2} \int \left[e^{-1} \dot{x}^\mu \dot{x}_\mu - e m^2 c^2\right] d\lambda \tag{9.6}$$

where $e(\lambda)$ is an auxiliary "einbein" field. Varying with respect to e gives:

$$e = \frac{\sqrt{\dot{x}^\mu \dot{x}_\mu}}{mc} \tag{9.7}$$

Substituting back recovers the original action.

9.1.5 Non-relativistic Limit

For $v \ll c$:

$$\sqrt{1 - v^2/c^2} \approx 1 - \frac{v^2}{2c^2} \tag{9.8}$$

So:

$$S \approx -mc^2 \int dt + \frac{1}{2} \int m v^2\, dt \tag{9.9}$$

The first term is a constant that doesn't affect the equations of motion. The second term is the familiar Newtonian kinetic energy.

9.1.6 Equations of Motion

Varying the action $S = -mc \int \sqrt{\dot{x}^\mu \dot{x}_\mu}\, d\lambda$:

$$\frac{d}{d\lambda}\left(\frac{\partial L}{\partial \dot{x}^\mu}\right) = 0 \tag{9.10}$$

where $L = -mc\sqrt{\dot{x}^\nu \dot{x}_\nu}$.

This gives:

$$\frac{d}{d\lambda}\left(\frac{\dot{x}_\mu}{\sqrt{\dot{x}^\nu \dot{x}_\nu}}\right) = 0 \tag{9.11}$$

Choosing proper time parametrization ($d\tau = ds/c$):

$$\frac{du_\mu}{d\tau} = 0 \tag{9.12}$$

where $u_\mu = dx_\mu/d\tau$ is the four-velocity. This is the equation for a straight worldline in Minkowski space.

9.2 Relativistic Action for a Free Particle

9.2.1 Four-Velocity and Four-Momentum

The four-velocity is:

$$u^\mu = \frac{dx^\mu}{d\tau} = \gamma(c, \mathbf{v}) \tag{9.13}$$

where $\gamma = 1/\sqrt{1 - v^2/c^2}$.

It satisfies the normalization:

$$u^\mu u_\mu = c^2 \tag{9.14}$$

The four-momentum is:

$$p^\mu = mu^\mu = (E/c, \mathbf{p}) \tag{9.15}$$

where:

$$E = \gamma mc^2 \quad \text{(relativistic energy)} \tag{9.16}$$

$$\mathbf{p} = \gamma m\mathbf{v} \quad \text{(relativistic momentum)} \tag{9.17}$$

9.2.2 Conservation Laws

From the free particle action, Noether's theorem gives:

- **Translation invariance**: $p^\mu = \text{constant}$
  - Energy conservation: $E = \text{constant}$
  - Momentum conservation: $\mathbf{p} = \text{constant}$
- **Lorentz invariance**: Angular momentum tensor $M^{\mu\nu} = x^\mu p^\nu - x^\nu p^\mu = \text{constant}$

9.2.3 The Energy-Momentum Relation

From $p^\mu p_\mu = m^2c^2$:

$$E^2 = p^2c^2 + m^2c^4 \tag{9.18}$$

This fundamental relation:

- Allows $E \neq 0$ even when $\mathbf{p} = 0$ (rest energy)
- Permits massless particles with $E = pc$
- Prevents $v > c$ for massive particles

9.2.4 Interaction with Electromagnetic Fields

A charged particle in an electromagnetic field has action:

$$S = \int \left[-mc\sqrt{\dot{x}^\mu \dot{x}_\mu} - qA_\mu \dot{x}^\mu\right] d\lambda \tag{9.19}$$

In terms of coordinate time:

$$S = \int \left[-mc^2\sqrt{1 - v^2/c^2} - q(\phi - \mathbf{v} \cdot \mathbf{A})\right] dt \tag{9.20}$$

This is the relativistic generalization of the minimal coupling prescription.

9.2.5 Canonical Momentum

The canonical momentum differs from mechanical momentum:

$$\mathbf{P} = \frac{\partial L}{\partial \mathbf{v}} = \gamma m\mathbf{v} + q\mathbf{A} = \mathbf{p} + q\mathbf{A} \tag{9.21}$$

The Hamiltonian is:

$$H = \mathbf{P} \cdot \mathbf{v} - L = \sqrt{(c\mathbf{P} - qc\mathbf{A})^2 + m^2c^4} + q\phi \tag{9.22}$$

9.2.6 Manifestly Covariant Equations

The Euler-Lagrange equations give:

$$\frac{dp^\mu}{d\tau} = qF^{\mu\nu}u_\nu \tag{9.23}$$

This is the Lorentz force law in covariant form. In 3-vector notation:

$$\frac{d\mathbf{p}}{dt} = q(\mathbf{E} + \mathbf{v} \times \mathbf{B}) \tag{9.24}$$

$$\frac{dE}{dt} = q\mathbf{E} \cdot \mathbf{v} \tag{9.25}$$

9.3 Relativistic Fields

9.3.1 Field Theory in Minkowski Space

Fields in special relativity are functions on Minkowski spacetime:

- Scalar field: $\phi(x^\mu)$
- Vector field: $A^\mu(x^\mu)$
- Tensor field: $T^{\mu\nu}(x^\mu)$

The action is:

$$S = \int \mathcal{L}(\phi, \partial_\mu \phi)\, d^4x \tag{9.26}$$

where $\mathcal{L}$ is the Lagrangian density.

9.3.2 Lorentz Invariant Actions

The Lagrangian density must be a Lorentz scalar. Building blocks:

- ϕ^2 (scalar)
- $\partial_\mu \phi \partial^\mu \phi$ (scalar)
- $F_{\mu\nu}F^{\mu\nu}$ (scalar)
- $\varepsilon^{\mu\nu\rho\sigma}F_{\mu\nu}F_{\rho\sigma}$ (pseudoscalar)

Example Lagrangians:

- Klein-Gordon: $\mathcal{L} = \frac{1}{2}\partial_\mu\phi\partial^\mu\phi - \frac{1}{2}m^2\phi^2$
- Maxwell: $\mathcal{L} = -\frac{1}{4}F_{\mu\nu}F^{\mu\nu}$
- Dirac: $\mathcal{L} = \bar{\psi}(i\gamma^\mu\partial_\mu - m)\psi$

9.3.3 Energy-Momentum Tensor

For a Lorentz invariant theory, the energy-momentum tensor:

$$T^{\mu\nu} = \frac{\partial\mathcal{L}}{\partial(\partial_\mu\phi)}\partial^\nu\phi - g^{\mu\nu}\mathcal{L} \tag{9.27}$$

is conserved: $\partial_\mu T^{\mu\nu} = 0$.

For the Klein-Gordon field:

$$T^{\mu\nu} = \partial^\mu\phi\partial^\nu\phi - g^{\mu\nu}\left[\frac{1}{2}\partial_\rho\phi\partial^\rho\phi - \frac{1}{2}m^2\phi^2\right] \tag{9.28}$$

9.3.4 Field Propagation and Causality

The Klein-Gordon equation:

$$(\Box + m^2)\phi = 0 \tag{9.29}$$

has solutions:

$$\phi(x) = \int \left[a(\mathbf{k})e^{-ik\cdot x} + a^\dagger(\mathbf{k})e^{ik\cdot x}\right]\frac{d^3k}{(2\pi)^3} \tag{9.30}$$

where $k^\mu = (\omega/c, \mathbf{k})$ with $\omega = \sqrt{k^2c^2 + m^2c^4}$.

The retarded Green's function:

$$G_R(x - x') = \theta(t - t')\int \frac{\sin(\omega|t - t'|)}{\omega}e^{i\mathbf{k}\cdot(\mathbf{x}-\mathbf{x}')}\frac{d^3k}{(2\pi)^3} \tag{9.31}$$

vanishes outside the light cone, ensuring causality.

9.3.5 Interactions and Gauge Fields

Interactions are introduced through:

1. Direct coupling: $\mathcal{L}_{\text{int}} = g\phi^3$

2. Gauge coupling: $\partial_\mu \to D_\mu = \partial_\mu + iqA_\mu$

3. Yukawa coupling: $\mathcal{L}_{\text{int}} = g\bar{\psi}\psi\phi$

Gauge invariance requires:

- Local symmetry: $\phi \to e^{iq\Lambda(x)}\phi$
- Gauge field transformation: $A_\mu \to A_\mu - \partial_\mu \Lambda$
- Covariant derivative: $D_\mu\phi \to e^{iq\Lambda} D_\mu\phi$

9.3.6 Spontaneous Symmetry Breaking

The Higgs mechanism starts with:

$$\mathcal{L} = \frac{1}{2}\partial_\mu\phi\partial^\mu\phi - V(\phi) \tag{9.32}$$

where $V(\phi) = -\mu^2\phi^2 + \lambda\phi^4$ has a minimum at $\phi_0 \neq 0$.

Expanding $\phi = \phi_0 + \eta$:

- The field η has mass $m_\eta = \sqrt{2\mu^2}$
- Gauge bosons acquire mass through the Higgs mechanism
- Goldstone modes appear for broken continuous symmetries

9.4 Variational Principle in General Relativity

9.4.1 From Special to General Relativity

Special relativity works in inertial frames, but gravity requires accelerated frames. Einstein's insight: incorporate gravity through curved spacetime geometry.

The metric tensor $g_{\mu\nu}(x)$ replaces the Minkowski metric $\eta_{\mu\nu}$:

$$ds^2 = g_{\mu\nu}dx^\mu dx^\nu \tag{9.33}$$

Principle of equivalence: locally, physics in a gravitational field is indistinguishable from physics in an accelerated frame.

9.4.2 Covariant Derivatives and Curvature

In curved spacetime, ordinary derivatives don't transform tensorially. The covariant derivative:

$$\nabla_\mu V^\nu = \partial_\mu V^\nu + \Gamma^\nu_{\mu\rho} V^\rho \tag{9.34}$$

where the connection coefficients:

$$\Gamma^\rho_{\mu\nu} = \frac{1}{2} g^{\rho\sigma} (\partial_\mu g_{\sigma\nu} + \partial_\nu g_{\sigma\mu} - \partial_\sigma g_{\mu\nu}) \tag{9.35}$$

The Riemann curvature tensor:

$$R^\rho{}_{\sigma\mu\nu} = \partial_\mu \Gamma^\rho_{\nu\sigma} - \partial_\nu \Gamma^\rho_{\mu\sigma} + \Gamma^\rho_{\mu\lambda} \Gamma^\lambda_{\nu\sigma} - \Gamma^\rho_{\nu\lambda} \Gamma^\lambda_{\mu\sigma} \tag{9.36}$$

measures the failure of covariant derivatives to commute:

$$[\nabla_\mu, \nabla_\nu] V^\rho = R^\rho{}_{\sigma\mu\nu} V^\sigma \tag{9.37}$$

9.4.3 The Principle of General Covariance

Physical laws must take the same form in all coordinate systems:

- Replace $\eta_{\mu\nu} \to g_{\mu\nu}$
- Replace $\partial_\mu \to \nabla_\mu$
- Replace $d^4x \to \sqrt{-g} d^4x$

This "minimal coupling" prescription extends special relativistic theories to curved spacetime.

9.4.4 Matter Action in Curved Spacetime

For matter fields ϕ in curved spacetime:

$$S_{\text{matter}} = \int \mathcal{L}_{\text{matter}}[\phi, \nabla_\mu \phi; g_{\mu\nu}] \sqrt{-g}\, d^4x \tag{9.38}$$

Examples:

- Scalar field: $\mathcal{L} = -\frac{1}{2}[g^{\mu\nu} \nabla_\mu \phi \nabla_\nu \phi + m^2 \phi^2 + \xi R \phi^2]$
- Electromagnetic: $\mathcal{L} = -\frac{1}{4} F_{\mu\nu} F^{\mu\nu} \sqrt{-g}$

The energy-momentum tensor:

$$T_{\mu\nu} = -\frac{2}{\sqrt{-g}} \frac{\delta S_{\text{matter}}}{\delta g^{\mu\nu}} \tag{9.39}$$

9.4.5 The Gravitational Action

What action for the gravitational field itself? Requirements:

- Depends only on $g_{\mu\nu}$ and derivatives
- At most second derivatives (avoid ghosts)
- Lorentz invariant in local inertial frames

The unique answer (up to cosmological constant):

$$S_{\text{gravity}} = \frac{c^3}{16\pi G} \int (R - 2\Lambda)\sqrt{-g}\, d^4x \tag{9.40}$$

where $R = g^{\mu\nu} R_{\mu\nu}$ is the Ricci scalar.

9.4.6 Boundary Terms

The Einstein-Hilbert action contains second derivatives of $g_{\mu\nu}$, requiring boundary terms for a well-posed variational principle. The Gibbons-Hawking-York term:

$$S_{\text{GHY}} = \frac{c^3}{8\pi G} \int_{\partial M} K\sqrt{h}\, d^3x \tag{9.41}$$

where K is the trace of the extrinsic curvature and h is the induced metric on the boundary.

9.5 Einstein-Hilbert Action

9.5.1 The Total Action

The complete gravitational action is:

$$S = S_{\text{EH}} + S_{\text{matter}} + S_{\text{GHY}} \tag{9.42}$$

where:

$$S_{\text{EH}} = \frac{c^3}{16\pi G} \int_M (R - 2\Lambda)\sqrt{-g}\, d^4x \tag{9.43}$$

$$S_{\text{matter}} = \int_M \mathcal{L}_{\text{matter}}\sqrt{-g}\, d^4x \tag{9.44}$$

9.5.2 Variation with Respect to the Metric

Computing $\delta S/\delta g^{\mu\nu}$ requires:

$$\delta\sqrt{-g} = -\frac{1}{2}\sqrt{-g}g_{\mu\nu}\delta g^{\mu\nu} \tag{9.45}$$

$$\delta\Gamma^{\rho}_{\mu\nu} = \frac{1}{2}g^{\rho\sigma}(\nabla_{\mu}\delta g_{\sigma\nu} + \nabla_{\nu}\delta g_{\sigma\mu} - \nabla_{\sigma}\delta g_{\mu\nu}) \tag{9.46}$$

$$\delta R_{\mu\nu} = \nabla_{\rho}\delta\Gamma^{\rho}_{\mu\nu} - \nabla_{\nu}\delta\Gamma^{\rho}_{\mu\rho} \tag{9.47}$$

After lengthy calculation:

$$\frac{\delta S_{\text{EH}}}{\delta g^{\mu\nu}} = -\frac{c^3}{16\pi G}\sqrt{-g}(R_{\mu\nu} - \frac{1}{2}g_{\mu\nu}R + \Lambda g_{\mu\nu}) \tag{9.48}$$

9.5.3 Einstein's Field Equations

Setting $\delta S/\delta g^{\mu\nu} = 0$:

$$R_{\mu\nu} - \frac{1}{2}g_{\mu\nu}R + \Lambda g_{\mu\nu} = \frac{8\pi G}{c^4}T_{\mu\nu} \tag{9.49}$$

These are Einstein's field equations, relating spacetime curvature to energy-momentum content.

Taking the trace:

$$R - 2R + 4\Lambda = \frac{8\pi G}{c^4}T \tag{9.50}$$

$$R = -\frac{8\pi G}{c^4}T + 4\Lambda \tag{9.51}$$

So the field equations can be written:

$$R_{\mu\nu} = \frac{8\pi G}{c^4}\left(T_{\mu\nu} - \frac{1}{2}g_{\mu\nu}T\right) + \Lambda g_{\mu\nu} \tag{9.52}$$

9.5.4 Conservation and Contracted Bianchi Identity

The contracted Bianchi identity:

$$\nabla^{\mu}\left(R_{\mu\nu} - \frac{1}{2}g_{\mu\nu}R\right) = 0 \tag{9.53}$$

ensures consistency with energy-momentum conservation:

$$\nabla^\mu T_{\mu\nu} = 0 \tag{9.54}$$

This isn't imposed but follows automatically from diffeomorphism invariance.

9.5.5 The Newtonian Limit

For weak fields $g_{\mu\nu} = \eta_{\mu\nu} + h_{\mu\nu}$ with $|h_{\mu\nu}| \ll 1$:

$$R_{00} \approx \frac{1}{2}\nabla^2 h_{00} \tag{9.55}$$

$$T_{00} \approx \rho c^2 \quad \text{(matter density)} \tag{9.56}$$

Einstein's equation gives:

$$\nabla^2 h_{00} = \frac{8\pi G}{c^2}\rho \tag{9.57}$$

Comparing with Newtonian gravity $\nabla^2\Phi = 4\pi G\rho$:

$$h_{00} = \frac{2\Phi}{c^2} \tag{9.58}$$

This confirms the correspondence principle.

9.5.6 Cosmological Solutions

For a homogeneous, isotropic universe, the metric takes the FRW form:

$$ds^2 = c^2dt^2 - a^2(t)\left[\frac{dr^2}{1-kr^2} + r^2(d\theta^2 + \sin^2\theta d\phi^2)\right] \tag{9.59}$$

Einstein's equations yield the Friedmann equations:

$$\left(\frac{\dot{a}}{a}\right)^2 = \frac{8\pi G}{3}\rho - \frac{kc^2}{a^2} + \frac{\Lambda c^2}{3} \tag{9.60}$$

$$\frac{\ddot{a}}{a} = -\frac{4\pi G}{3}\left(\rho + \frac{3p}{c^2}\right) + \frac{\Lambda c^2}{3} \tag{9.61}$$

These govern cosmic expansion.

9.6 Geodesics and Principle of Least Action

9.6.1 Geodesic Action

A free particle in curved spacetime follows a geodesic—the straightest possible path. The action is:

$$S = -mc \int ds = -mc \int \sqrt{g_{\mu\nu}\dot{x}^\mu \dot{x}^\nu}\, d\lambda \tag{9.62}$$

This is the natural generalization of the special relativistic action.

9.6.2 The Geodesic Equation

Varying the action:

$$\delta S = -mc \int \frac{g_{\mu\nu}\dot{x}^\mu \delta\dot{x}^\nu}{\sqrt{g_{\rho\sigma}\dot{x}^\rho \dot{x}^\sigma}}\, d\lambda \tag{9.63}$$

Integration by parts and using $\delta\dot{x}^\mu = d(\delta x^\mu)/d\lambda$:

$$\frac{d^2x^\mu}{d\tau^2} + \Gamma^\mu_{\nu\rho}\frac{dx^\nu}{d\tau}\frac{dx^\rho}{d\tau} = 0 \tag{9.64}$$

This is the geodesic equation, with τ the proper time.

9.6.3 Physical Interpretation

The geodesic equation can be written:

$$\frac{du^\mu}{d\tau} + \Gamma^\mu_{\nu\rho}u^\nu u^\rho = 0 \tag{9.65}$$

or:

$$\nabla_u u = 0 \tag{9.66}$$

The four-acceleration vanishes in the covariant sense—particles follow straight lines in curved spacetime.

9.6.4 Gravitational Redshift

For a static metric, the timelike Killing vector $\xi^\mu = (1, 0, 0, 0)$ gives a conserved quantity:

$$E = -g_{\mu\nu}\xi^\mu p^\nu = g_{00}p^0 \tag{9.67}$$

For photons in the Schwarzschild metric:

$$g_{00} = 1 - \frac{2GM}{rc^2} \tag{9.68}$$

The frequency ratio:

$$\frac{\nu_\infty}{\nu_r} = \sqrt{g_{00}(r)} = \sqrt{1 - \frac{2GM}{rc^2}} \tag{9.69}$$

This is gravitational redshift.

9.6.5 Perihelion Precession

For planetary orbits in the Schwarzschild metric, the effective potential is:

$$V_{\text{eff}} = -\frac{GM}{r} + \frac{L^2}{2mr^2} - \frac{GML^2}{mc^2r^3} \tag{9.70}$$

The last term, absent in Newton's theory, causes perihelion precession:

$$\Delta\phi = \frac{6\pi GM}{c^2a(1 - e^2)} \tag{9.71}$$

per orbit, where a is the semi-major axis and e the eccentricity.

9.6.6 Fermat's Principle in Curved Spacetime

Light rays follow null geodesics: $ds^2 = 0$. The action:

$$S = \int \sqrt{g_{ij}dx^i dx^j} \tag{9.72}$$

where the integral is over the spatial path.

This is Fermat's principle in curved spacetime—light takes the path of extremal optical length. Gravitational lensing follows directly.

Summary

The principle of least action not only survives the transition to relativity but becomes even more fundamental. In special relativity, the requirement of Lorentz invariance uniquely determines the form of the action, automatically preventing superluminal motion and yielding the correct energy-momentum relations. The action principle provides the natural framework for relativistic field theories, ensuring causal propagation and proper transformation properties.

In general relativity, the principle reaches its zenith. Gravity itself emerges from an action principle—the Einstein-Hilbert action—transforming from a mysterious force to the curvature of spacetime. The variational approach:

1. **Unifies Geometry and Dynamics**: Einstein's equations relate the geometry of spacetime (left side) to its matter-energy content (right side), both derived from a single action.

2. **Ensures Consistency**: The contracted Bianchi identity, following from diffeomorphism invariance, automatically ensures energy-momentum conservation.

3. **Reveals Physical Content**: Symmetries of the action yield conserved quantities even in curved spacetime, while the equivalence principle emerges naturally.

4. **Guides Quantum Extensions**: The action formulation provides the starting point for quantum gravity attempts, from perturbative quantum general relativity to string theory.

5. **Predicts New Phenomena**: Gravitational waves, black holes, and cosmological expansion all follow from the Einstein-Hilbert action.

The success of the action principle in relativity demonstrates its fundamental role in physics. Far from being a mere reformulation of Newton's laws, it provides the key to understanding spacetime itself. As we'll see in subsequent chapters, this same principle bridges classical and quantum physics, offering our best hope for a unified understanding of nature.

Chapter 10

Quantum Mechanics

The transition from classical to quantum mechanics represents one of the most profound conceptual shifts in physics. Remarkably, the principle of least action not only survives this transition but provides the crucial bridge between classical and quantum worlds. Through Feynman's path integral formulation, we discover that quantum mechanics can be understood as a theory where particles explore all possible paths, with the classical path emerging from quantum interference. The action, which determines a unique classical trajectory, now assigns probability amplitudes to all trajectories.

This chapter explores how the variational principle manifests in quantum mechanics, revealing deep connections between classical action and quantum amplitudes. We'll see how the Schrödinger equation emerges as a limiting case of the path integral, how the classical limit is recovered through stationary phase approximation, and how semiclassical methods bridge the quantum-classical divide. From the hydrogen atom to quantum tunneling, from molecular vibrations to quantum chaos, the action principle illuminates quantum phenomena across all scales.

10.1 Feynman Path Integral

10.1.1 The Revolutionary Idea

In 1948, Richard Feynman introduced a formulation of quantum mechanics that seems to contradict classical physics while simultaneously explaining it. His key insight: a quantum particle doesn't follow a single path but explores all possible paths simultaneously.

The probability amplitude for a particle to go from point (x_a, t_a) to point (x_b, t_b) is:

$$K(x_b, t_b; x_a, t_a) = \sum_{\text{all paths}} e^{iS[\text{path}]/\hbar} \tag{10.1}$$

where $S[\text{path}]$ is the classical action along that path.

This immediately raises questions:

- How do we sum over all paths?
- Why does the classical action appear?
- How does this reproduce standard quantum mechanics?

10.1.2 Mathematical Formulation

To make the path integral precise, we discretize time:

- Divide interval $[t_a, t_b]$ into N segments: $\varepsilon = (t_b - t_a)/N$
- Path specified by positions: $x_0 = x_a, x_1, x_2, ..., x_N = x_b$

The propagator becomes:

$$K = \lim_{N\to\infty} \int \cdots \int \left(\frac{m}{2\pi i\hbar\varepsilon}\right)^{N/2} \exp\left[\frac{i}{\hbar}\sum_{j=0}^{N-1} L\left(x_j, \frac{x_{j+1} - x_j}{\varepsilon}, t_j\right)\varepsilon\right] dx_1...dx_{N-1} \tag{10.2}$$

where L is the classical Lagrangian.

10.1.3 Free Particle Path Integral

For a free particle, $L = \frac{1}{2}m\left(\frac{dx}{dt}\right)^2$:

$$S = \int_{t_a}^{t_b} \frac{m}{2}\left(\frac{dx}{dt}\right)^2 dt \tag{10.3}$$

For the discretized path:

$$S \approx \sum_{j=0}^{N-1} \frac{m}{2\varepsilon}(x_{j+1} - x_j)^2 \tag{10.4}$$

The path integral becomes Gaussian:

$$K = \lim_{N\to\infty} \left(\frac{m}{2\pi i\hbar\varepsilon}\right)^{N/2} \int \cdots \int \exp\left[\frac{im}{2\hbar\varepsilon}\sum(x_{j+1}-x_j)^2\right] dx_1 ... dx_{N-1} \tag{10.5}$$

10.1.4 Evaluation of the Free Particle Propagator

Each integral is Gaussian. Using:

$$\int_{-\infty}^{\infty} \exp[-(a+ib)x^2]dx = \sqrt{\frac{\pi}{a+ib}} \tag{10.6}$$

Sequential integration yields:

$$K(x_b, t_b; x_a, t_a) = \sqrt{\frac{m}{2\pi i\hbar(t_b - t_a)}} \exp\left[\frac{im(x_b - x_a)^2}{2\hbar(t_b - t_a)}\right] \tag{10.7}$$

This is the exact quantum mechanical propagator for a free particle.

10.1.5 Connection to Wave Functions

The propagator relates wave functions at different times:

$$\psi(x_b, t_b) = \int K(x_b, t_b; x_a, t_a)\psi(x_a, t_a)dx_a \tag{10.8}$$

This integral equation is equivalent to the Schrödinger equation:

$$i\hbar\frac{\partial\psi}{\partial t} = \hat{H}\psi \tag{10.9}$$

The path integral thus provides a complete formulation of quantum mechanics.

10.1.6 Physical Interpretation

The path integral embodies wave-particle duality:

- Particle aspect: We consider trajectories $x(t)$
- Wave aspect: Amplitudes $e^{iS/\hbar}$ interfere

Key features:

- All paths contribute with equal magnitude $|e^{iS/\hbar}| = 1$
- Phase $S/\hbar$ determines interference
- Classical path emerges from stationary phase condition $\delta S = 0$

10.2 Quantum Principle of Least Action

10.2.1 From Extremal to Stationary Phase

In classical mechanics, the particle follows the path where $\delta S = 0$. In quantum mechanics, all paths contribute, but paths where S varies rapidly interfere destructively.

Near the classical path $x_{cl}(t)$:

$$S[x_{cl} + \delta x] \approx S[x_{cl}] + \frac{1}{2}\int\int \delta x(t) \frac{\delta^2 S}{\delta x(t)\delta x(t')} \delta x(t') dt dt' \tag{10.10}$$

Since $\delta S = 0$ at x_{cl}, the linear term vanishes.

10.2.2 The Principle of Stationary Phase

For $\hbar \to 0$, the phase $S/\hbar$ oscillates rapidly except near stationary points. The dominant contribution comes from paths where:

$$\frac{\delta S[x]}{\delta x(t)} = 0 \tag{10.11}$$

This yields the Euler-Lagrange equation—the classical equation of motion emerges from quantum mechanics!

10.2.3 Quantum Fluctuations

The quadratic term in the action expansion determines quantum fluctuations:

$$K \approx \exp(iS_{cl}/\hbar) \times \int \mathcal{D}[\delta x] \exp\left[\frac{i}{2\hbar}\int\int \delta x \cdot \left(\frac{\delta^2 S}{\delta x \delta x}\right) \cdot \delta x\right] \tag{10.12}$$

The functional integral over fluctuations δx gives:

$$K = \exp(iS_{cl}/\hbar) \times \left[\text{Det}\left(\frac{\delta^2 S}{\delta x \delta x}\right)\right]^{-1/2} \tag{10.13}$$

This determinant encodes quantum corrections.

10.2.4 The Classical Action as a Phase

The appearance of $S/\hbar$ as a phase has profound implications:

- Action differences of order $\hbar$ produce significant phase differences
- Classical mechanics emerges when $S \gg \hbar$
- Quantum effects dominate when $S \sim \hbar$

This explains why quantum mechanics is important for:

- Small masses (small action)
- Short times (small action)
- Low energies (small action differences)

10.2.5 Topological Effects

In multiply-connected spaces, paths fall into topological classes. The action can differ by:

$$\Delta S = \oint p \cdot dx = nh \quad \text{(Bohr-Sommerfeld)} \tag{10.14}$$

This leads to:

- Quantization conditions
- Aharonov-Bohm effect
- Instantons in field theory

10.2.6 The Quantum Action Principle

Schwinger formulated quantum mechanics through:

$$\delta\langle x_b, t_b|x_a, t_a\rangle = \frac{i}{\hbar}\langle x_b, t_b|\delta S|x_a, t_a\rangle \tag{10.15}$$

This quantum action principle:

- Generates equations of motion
- Yields commutation relations
- Provides systematic approximation schemes

10.3 Analogy between Optics and Quantum Mechanics

10.3.1 Historical Connections

The wave-particle duality in quantum mechanics mirrors the ray-wave duality in optics:

Optics	**Quantum Mechanics**
Ray optics	Classical mechanics
Wave optics	Quantum mechanics
Wavelength λ	de Broglie wavelength $\lambda = h/p$
Fermat's principle	Principle of least action
Eikonal	Classical action

10.3.2 The Eikonal Approximation

In optics, for short wavelengths, the wave equation:

$$\nabla^2\psi + k^2\psi = 0 \tag{10.16}$$

has solutions:

$$\psi = A(\mathbf{r})\exp[iS(\mathbf{r})/\hbar] \tag{10.17}$$

where S satisfies the eikonal equation:

$$|\nabla S|^2 = n^2(\mathbf{r})p^2 \tag{10.18}$$

with $n(\mathbf{r})$ the refractive index.

10.3.3 Hamilton-Jacobi and Schrödinger

The time-independent Schrödinger equation:

$$-\frac{\hbar^2}{2m}\nabla^2\psi + V\psi = E\psi \tag{10.19}$$

With $\psi = \exp[iS/\hbar]$:

$$|\nabla S|^2 = 2m(E - V) + \frac{i\hbar}{2}\nabla^2 S \tag{10.20}$$

The classical limit $\hbar \to 0$ gives the Hamilton-Jacobi equation:

$$|\nabla S|^2 = 2m(E - V) \tag{10.21}$$

10.3.4 Quantum Trajectories

Defining the quantum potential:

$$Q = -\frac{\hbar^2}{2m}\frac{\nabla^2\sqrt{\rho}}{\sqrt{\rho}} \tag{10.22}$$

where $\rho = |\psi|^2$, the Hamilton-Jacobi equation becomes:

$$|\nabla S|^2 = 2m(E - V - Q) \tag{10.23}$$

This suggests quantum particles follow classical trajectories in a modified potential $V + Q$.

10.3.5 Interference and Diffraction

Wave phenomena in optics have quantum analogs:

Double-slit interference:

- Optics: $I = |E_1 + E_2|^2$
- Quantum: $P = |\psi_1 + \psi_2|^2$

Diffraction:

- Optics: Huygens' principle
- Quantum: Path integral over all paths

Coherence:

- Optics: Phase relationships
- Quantum: Off-diagonal density matrix elements

10.3.6 Berry Phase and Geometric Optics

The Berry phase in quantum mechanics:

$$\gamma = i \oint \langle \psi | \nabla \psi \rangle \cdot d\mathbf{r} \tag{10.24}$$

parallels the geometric phase in optics:

$$\phi = \oint \mathbf{k} \cdot d\mathbf{r} \tag{10.25}$$

Both arise from parallel transport in appropriate spaces.

10.4 Schrödinger Equation as a Limit

10.4.1 Deriving Schrödinger from Path Integrals

For infinitesimal time ε:

$$\psi(x, t+\varepsilon) = \int K(x, t+\varepsilon; x', t)\psi(x', t)dx' \tag{10.26}$$

The short-time propagator:

$$K(x, t+\varepsilon; x', t) \approx \left(\frac{m}{2\pi i\hbar\varepsilon}\right)^{1/2} \exp\left[\frac{i}{\hbar} L\left(\frac{x+x'}{2}, \frac{x-x'}{\varepsilon}, t\right)\varepsilon\right] \tag{10.27}$$

10.4.2 Taylor Expansion

Expanding for small ε and $(x - x')$:

$$\psi(x, t+\varepsilon) \approx \psi(x, t) + \varepsilon\frac{\partial\psi}{\partial t} \tag{10.28}$$

$$K \approx \left(\frac{m}{2\pi i\hbar\varepsilon}\right)^{1/2} \exp\left[\frac{im(x-x')^2}{2\hbar\varepsilon} - \frac{iV\varepsilon}{\hbar}\right] \tag{10.29}$$

10.4.3 Gaussian Integration

Performing the Gaussian integral over x':

$$\int \exp\left[\frac{im(x-x')^2}{2\hbar\varepsilon}\right] f(x')dx' \approx f(x) + \frac{i\hbar\varepsilon}{2m}\nabla^2 f(x) \tag{10.30}$$

Combining terms:

$$\psi(x, t) + \varepsilon\frac{\partial\psi}{\partial t} = \left[1 - \frac{iV\varepsilon}{\hbar}\right]\left[\psi(x, t) + \frac{i\hbar\varepsilon}{2m}\nabla^2\psi\right] \tag{10.31}$$

10.4.4 The Schrödinger Equation

Keeping terms to order ε:

$$\frac{\partial\psi}{\partial t} = -\frac{iV}{\hbar}\psi + \frac{i\hbar}{2m}\nabla^2\psi \tag{10.32}$$

Rearranging:

$$i\hbar\frac{\partial\psi}{\partial t} = -\frac{\hbar^2}{2m}\nabla^2\psi + V\psi \tag{10.33}$$

This is the Schrödinger equation!

10.4.5 Operator Correspondence

The path integral naturally yields operator correspondences:

$$\text{Position: } \hat{x} = x \tag{10.34}$$

$$\text{Momentum: } \hat{p} = -i\hbar\nabla \tag{10.35}$$

$$\text{Hamiltonian: } \hat{H} = \frac{\hat{p}^2}{2m} + V(\hat{x}) \tag{10.36}$$

These follow from the short-time propagator structure.

10.4.6 Time Evolution Operator

The finite-time propagator:

$$K(x_b, t_b; x_a, t_a) = \langle x_b| \exp\left[-i\hat{H}(t_b - t_a)/\hbar\right] |x_a\rangle \tag{10.37}$$

provides the time evolution operator:

$$\hat{U}(t) = \exp\left(-i\hat{H}t/\hbar\right) \tag{10.38}$$

The path integral thus gives a constructive definition of this operator.

10.5 Semiclassical Approximation

10.5.1 The WKB Method

When $S \gg \hbar$, we can expand systematically in powers of $\hbar$:

$$\psi = \exp\left[\frac{i}{\hbar}(S_0 + \hbar S_1 + \hbar^2 S_2 + ...)\right] \tag{10.39}$$

Substituting into the Schrödinger equation and collecting powers of $\hbar$:

$$\hbar^0: \quad |\nabla S_0|^2 = 2m(E - V) \quad \text{(Hamilton-Jacobi)} \tag{10.40}$$

$$\hbar^1: \quad \nabla^2 S_0 + 2\nabla S_0 \cdot \nabla S_1 = 0 \quad \text{(Transport equation)} \tag{10.41}$$

10.5.2 WKB Wave Functions

The leading-order solution:

$$\psi_{WKB} = \frac{A}{\sqrt{p}} \exp\left[\pm \frac{i}{\hbar} \int p dx\right] \tag{10.42}$$

where $p = \sqrt{2m(E - V)}$.

Connection formulas at turning points where $p = 0$:

- Decaying to oscillating: Airy function matching
- Phase shift of $\pi/4$

10.5.3 Bohr-Sommerfeld Quantization

For bound states, the wave function must be single-valued:

$$\oint p dx = 2\pi\hbar \left(n + \frac{\mu}{4}\right) \tag{10.43}$$

where μ is the Maslov index counting turning points.

This yields energy levels:

- Harmonic oscillator: $E_n = \hbar\omega(n + 1/2)$
- Hydrogen atom: $E_n = -13.6 \text{ eV}/n^2$

10.5.4 Tunneling and Instantons

For $E < V$, the classical momentum becomes imaginary:

$$p = i|p| = i\sqrt{2m(V - E)} \tag{10.44}$$

The WKB tunneling probability:

$$T = \exp\left[-2 \int \frac{|p|dx}{\hbar}\right] = \exp[-2S_E/\hbar] \tag{10.45}$$

where S_E is the Euclidean action of the instanton.

10.5.5 Path Integral in the Semiclassical Limit

The path integral is dominated by paths near the classical path:

$$K \approx \sum_{\text{classical paths}} A_j \exp[iS_j/\hbar] \tag{10.46}$$

where:

- S_j = classical action along path j
- $A_j = (2\pi i\hbar)^{-f/2} \left[\text{Det}\left(\frac{\partial^2 S_j}{\partial x_a \partial x_b}\right)\right]^{1/2}$
- f = number of conjugate points

Multiple classical paths lead to interference.

10.5.6 Applications of Semiclassical Methods

- **Molecular spectroscopy:** Franck-Condon factors from classical trajectories
- **Quantum chaos:** Gutzwiller trace formula:

$$\rho(E) \approx \bar{\rho}(E) + \sum_{\text{periodic orbits}} A_p \cos(S_p/\hbar) \tag{10.47}$$

- **Mesoscopic physics:** Conductance fluctuations from interfering paths
- **Chemical dynamics:** Reaction rates from instanton calculations

10.6 Summary

The path integral formulation reveals the principle of least action as the fundamental bridge between classical and quantum mechanics. Rather than abandoning the action in quantum theory, we discover it plays an even more central role—determining not which path is taken, but the probability amplitude for every possible path.

Key insights from this quantum perspective:

1. **Superposition of Histories:** Quantum mechanics includes all possible classical histories, weighted by $\exp(iS/\hbar)$. The classical world emerges not from selecting one history but from the constructive interference of many.

2. **Action as Phase:** The classical action becomes a quantum mechanical phase. This explains why $\hbar$ sets the scale for quantum phenomena—when action changes are comparable to $\hbar$, quantum effects dominate.

3. **Classical Limit:** The principle of stationary phase shows that classical paths emerge naturally from quantum mechanics when $S \gg \hbar$. Classical mechanics is not replaced but encompassed.

4. **Unification of Formulations:** The path integral unifies different approaches:
   - Yields Schrödinger equation in the infinitesimal limit
   - Reduces to Hamilton-Jacobi in the classical limit
   - Connects to matrix mechanics through operator insertions

5. **Computational Power:** Semiclassical methods based on the action provide:
   - Quantization conditions (Bohr-Sommerfeld)
   - Tunneling rates (instantons)
   - Quantum corrections (loop expansion)

The success of the path integral approach extends far beyond quantum mechanics. It provides the foundation for quantum field theory, where the sum over paths becomes a sum over field configurations. Statistical mechanics emerges from Euclidean path integrals. Even quantum gravity attempts start from path integrals over metrics.

The principle of least action thus reveals itself not as a classical curiosity but as the fundamental organizing principle of quantum physics. The action determines not what will happen but what can happen—and with what probability amplitude. In this quantum view, the classical principle of least action emerges as the saddle point of a far richer quantum reality.

Chapter 11

Quantum Field Theory

Quantum field theory (QFT) represents the synthesis of quantum mechanics and special relativity, providing our most successful framework for understanding fundamental particles and their interactions. The principle of least action plays an even more central role in QFT than in quantum mechanics—the entire theory is constructed from actions that respect both quantum principles and relativistic symmetries. From the electromagnetic force to the strong and weak nuclear forces, all fundamental interactions are described by action functionals.

This chapter explores how the path integral extends from particle quantum mechanics to field theory, where we sum over all possible field configurations rather than particle paths. We'll see how Feynman diagrams emerge naturally from expanding the action, how symmetries lead to conserved currents and selection rules, and how the process of renormalization tames the infinities that arise. The principle of least action not only survives the transition to quantum fields but becomes the organizing principle for all of particle physics.

11.1 Action for Quantized Fields

11.1.1 From Particles to Fields

In quantum mechanics, we sum over particle paths $x(t)$. In quantum field theory, we sum over field configurations $\phi(x, t)$. The transition is:

$$\text{QM path integral:} \quad \int \mathcal{D}[x(t)] \exp\{iS[x]/\hbar\} \tag{11.1}$$

$$\text{QFT path integral:} \quad \int \mathcal{D}[\phi(x,t)] \exp\{iS[\phi]/\hbar\} \tag{11.2}$$

where $S[\phi] = \int d^4x \mathcal{L}[\phi, \partial_\mu \phi]$ is the action functional.

The field $\phi(x,t)$ has a value at each spacetime point—infinitely many degrees of freedom. This allows:

- Particle creation and annihilation
- Lorentz invariant description
- Natural incorporation of antiparticles

11.1.2 Canonical Quantization vs Path Integrals

Two approaches to quantizing fields:

Canonical Quantization:

- Promote fields to operators: $\phi(x) \to \hat{\phi}(x)$
- Impose commutation relations: $[\hat{\phi}(x,t), \hat{\pi}(y,t)] = i\hbar\delta^3(x-y)$
- Build Fock space of multi-particle states

Path Integral Quantization:

- Sum over classical field configurations
- Weight by $\exp(iS/\hbar)$
- Generate correlation functions directly

The path integral approach:

- Makes Lorentz invariance manifest
- Handles gauge theories naturally
- Connects to statistical mechanics
- Provides non-perturbative insights

11.1.3 Free Field Path Integral

For a free scalar field with action:

$$S = \int d^4x \left[\frac{1}{2}\partial_\mu\phi\partial^\mu\phi - \frac{1}{2}m^2\phi^2\right] \tag{11.3}$$

The path integral is Gaussian:

$$Z[J] = \int \mathcal{D}[\phi] \exp\left\{i \int d^4x \left[\frac{1}{2}\partial_\mu\phi\partial^\mu\phi - \frac{1}{2}m^2\phi^2 + J\phi\right] /\hbar\right\} \tag{11.4}$$

Completing the square:

$$Z[J] = \mathcal{N} \exp\left\{-\frac{i}{2\hbar}\int d^4x d^4y J(x)G(x-y)J(y)\right\} \tag{11.5}$$

where $G(x-y)$ is the Feynman propagator:

$$(\Box + m^2)G(x-y) = -i\delta^4(x-y) \tag{11.6}$$

11.1.4 The Feynman Propagator

In momentum space:

$$\tilde{G}(p) = \frac{i}{p^2 - m^2 + i\varepsilon} \tag{11.7}$$

The $i\varepsilon$ prescription ensures:

- Positive energy particles propagate forward in time
- Negative energy particles propagate backward (antiparticles)
- Causality is maintained

Position space propagator:

$$G(x-y) = \int \frac{d^4p}{(2\pi)^4}\frac{i}{p^2 - m^2 + i\varepsilon}e^{-ip\cdot(x-y)} \tag{11.8}$$

This is the amplitude for a particle to propagate from y to x.

11.1.5 Interacting Fields

For interacting theories, add interaction terms:

$$S = S_0 + S_{int} \tag{11.9}$$

Example - ϕ^4 theory:

$$S = \int d^4x \left[\frac{1}{2}\partial_\mu\phi\partial^\mu\phi - \frac{1}{2}m^2\phi^2 - \frac{\lambda}{4!}\phi^4\right] \tag{11.10}$$

The path integral becomes:

$$Z[J] = \int \mathcal{D}[\phi] \exp\{i[S_0[\phi] + S_{int}[\phi] + \int J\phi]/\hbar\} \tag{11.11}$$

This cannot be evaluated exactly, requiring perturbation theory.

11.1.6 Generating Functionals

The partition function $Z[J]$ generates correlation functions:

$$\langle\phi(x_1)...\phi(x_n)\rangle = \frac{1}{Z[0]} \times \left.\frac{\delta^n Z[J]}{\delta J(x_1)...\delta J(x_n)}\right|_{J=0} \tag{11.12}$$

Connected correlations come from $W[J] = -i\hbar \ln Z[J]$:

$$\langle\phi(x_1)...\phi(x_n)\rangle_c = \left.\frac{\delta^n W[J]}{\delta J(x_1)...\delta J(x_n)}\right|_{J=0} \tag{11.13}$$

The effective action $\Gamma[\phi_c]$ is the Legendre transform of $W[J]$:

$$\Gamma[\phi_c] = W[J] - \int J(x)\phi_c(x)d^4x \tag{11.14}$$

where $\phi_c = \delta W/\delta J$ is the classical field.

11.2 Symmetries and Noether's Theorem in QFT

11.2.1 Continuous Symmetries in Field Theory

A field theory has a continuous symmetry if the action is invariant under:

$$\phi(x) \to \phi'(x) = \phi(x) + \varepsilon\Delta\phi(x) \tag{11.15}$$

where ε is infinitesimal.

For the action to be invariant:

$$\delta S = \int d^4x \left(\frac{\partial \mathcal{L}}{\partial \phi} \Delta\phi + \frac{\partial \mathcal{L}}{\partial(\partial_\mu \phi)} \partial_\mu(\Delta\phi) \right) = 0 \tag{11.16}$$

11.2.2 Noether's Theorem for Fields

Using the Euler-Lagrange equations:

$$\frac{\partial \mathcal{L}}{\partial \phi} - \partial_\mu \left(\frac{\partial \mathcal{L}}{\partial(\partial_\mu \phi)} \right) = 0 \tag{11.17}$$

We find:

$$\partial_\mu j^\mu = 0 \tag{11.18}$$

where the Noether current is:

$$j^\mu = \frac{\partial \mathcal{L}}{\partial(\partial_\mu \phi)} \Delta\phi - K^\mu \tag{11.19}$$

K^μ appears for spacetime symmetries.

11.2.3 Examples of Conserved Currents

Global U(1) symmetry: $\phi \to e^{i\alpha}\phi$

- Current: $j^\mu = i(\phi^* \partial^\mu \phi - \phi \partial^\mu \phi^*)$
- Conserved charge: $Q = \int j^0 d^3x$ (particle number)

Translation invariance: $x^\mu \to x^\mu + a^\mu$

- Energy-momentum tensor: $T^{\mu\nu} = \frac{\partial \mathcal{L}}{\partial(\partial_\mu \phi)} \partial^\nu \phi - g^{\mu\nu} \mathcal{L}$
- Conserved quantities: $P^\mu = \int T^{0\mu} d^3x$

Lorentz invariance: $x^\mu \to \Lambda^\mu_\nu x^\nu$

- Angular momentum tensor: $M^{\mu\nu\rho} = x^\nu T^{\mu\rho} - x^\rho T^{\mu\nu} + S^{\mu\nu\rho}$
- Conserved quantities: $J^{\mu\nu} = \int M^{0\mu\nu} d^3x$

11.2.4 Ward-Takahashi Identities

In the quantum theory, symmetries lead to relations between correlation functions. For a symmetry with current j^μ:

$$\langle \partial_\mu j^\mu(x)\phi(x_1)...\phi(x_n)\rangle = -i\sum_k \delta^4(x-x_k)\langle\phi(x_1)...\Delta\phi(x_k)...\phi(x_n)\rangle \tag{11.20}$$

These Ward-Takahashi identities:

- Constrain scattering amplitudes
- Ensure gauge invariance
- Relate different Green's functions
- Protect massless particles

11.2.5 Spontaneous Symmetry Breaking

When $\langle\phi\rangle \neq 0$, a symmetric action can have non-symmetric ground state:

$$\mathcal{L} = \frac{1}{2}\partial_\mu\phi\partial^\mu\phi + \mu^2\phi^2 - \lambda\phi^4 \tag{11.21}$$

For $\mu^2 < 0$, the minimum is at $\phi = \pm v$ where $v = \sqrt{-\mu^2/2\lambda}$. Expanding $\phi = v + \eta$:

- The field η has mass $m_\eta = 2\sqrt{-\mu^2}$
- Goldstone theorem: Broken continuous symmetries $\rightarrow$ massless bosons
- In gauge theories: Goldstone bosons become longitudinal gauge boson components

11.2.6 Anomalies

Some classical symmetries fail in the quantum theory due to regularization requirements. The chiral anomaly:

$$\partial_\mu j^{\mu 5} = \frac{e^2}{16\pi^2}\tilde{F}^{\mu\nu}F_{\mu\nu} \tag{11.22}$$

where $j^{\mu 5} = \bar{\psi}\gamma^\mu\gamma^5\psi$ is the axial current.

Consequences:

- $\pi^0 \to 2\gamma$ decay
- Constraints on fermion content (anomaly cancellation)
- Topological effects in gauge theories

11.3 Feynman Rules from the Action

11.3.1 Perturbation Theory

For $S = S_0 + S_{int}$, expand the path integral:

$$Z[J] = \int \mathcal{D}[\phi] e^{iS_0/\hbar} \left[1 + \frac{iS_{int}}{\hbar} + \frac{(i/\hbar)^2 S_{int}^2}{2!} + ...\right] e^{i\int J\phi/\hbar} \tag{11.23}$$

Each term involves Gaussian integrals that can be evaluated using Wick's theorem.

11.3.2 Wick's Theorem

For free fields, all correlation functions decompose into products of propagators:

$$\langle T[\phi(x_1)...\phi(x_{2n})]\rangle_0 = \sum_{\text{pairings}} \prod_{\text{pairs}} G(x_i - x_j) \tag{11.24}$$

Time ordering T ensures causality.
Example:

$$\langle T[\phi(x_1)\phi(x_2)\phi(x_3)\phi(x_4)]\rangle_0 = G(x_1-x_2)G(x_3-x_4)+G(x_1-x_3)G(x_2-x_4)+G(x_1-x \tag{11.25}$$

11.3.3 Feynman Diagrams

Each term in perturbation theory corresponds to a Feynman diagram:

- **Propagators:** Lines representing $G(x - y)$
  - Scalar: —
  - Fermion: →—
  - Photon: ∼∼∼

- **Vertices:** Interaction points from S_{int}
  - ϕ^4: Four lines meeting
  - QED: Photon-electron-positron vertex
- **External legs:** Sources $J(x)$

11.3.4 Momentum Space Rules

After Fourier transformation, Feynman rules in momentum space:

ϕ^4 **theory:**

- Propagator: $\frac{i}{p^2-m^2+i\varepsilon}$
- Vertex: $-i\lambda$
- Integration: $\int \frac{d^4k}{(2\pi)^4}$ for each loop
- Conservation: $(2\pi)^4\delta^4(\sum p)$ at each vertex

QED:

- Electron propagator: $\frac{i(\not{p}+m)}{p^2-m^2+i\varepsilon}$
- Photon propagator: $\frac{-ig_{\mu\nu}}{k^2}$
- Vertex: $-ie\gamma^\mu$

11.3.5 S-Matrix Elements

The S-matrix element for $i \to f$:

$$S_{fi} = \langle f|S|i\rangle = \delta_{fi} + i(2\pi)^4\delta^4(p_f - p_i)\mathcal{M}_{fi} \tag{11.26}$$

where $\mathcal{M}_{fi}$ is the invariant amplitude calculated from:

- Connected diagrams only
- Amputated external propagators
- Sum over all topologies at each order

Cross sections:

$$d\sigma = \frac{|\mathcal{M}_{fi}|^2 d\Phi}{\text{flux}} \tag{11.27}$$

where $d\Phi$ is the phase space element.

11.3.6 Loop Integrals and Regularization

Loop integrals often diverge:

$$\int \frac{d^4k}{k^2} \to \infty \tag{11.28}$$

Regularization methods:

- Pauli-Villars: Add heavy fictitious particles
- Dimensional regularization: Work in $d = 4 - \varepsilon$ dimensions
- Cutoff: Integrate up to Λ

Example - one-loop self-energy in ϕ^4:

$$\Sigma(p^2) = \frac{\lambda}{2} \int \frac{d^4k}{(2\pi)^4(k^2 - m^2 + i\varepsilon)} \tag{11.29}$$

This diverges logarithmically, requiring renormalization.

11.4 Renormalization and Renormalization Group

11.4.1 The Problem of Infinities

Loop calculations yield infinities:

- Self-energy: $\Sigma(p^2) \sim \Lambda^2$ (mass renormalization)
- Vertex correction: $\Gamma \sim \ln \Lambda$ (coupling renormalization)
- Vacuum bubbles: $\sim \Lambda^4$ (cosmological constant)

Physical observables must be finite—how do we extract them?

11.4.2 Renormalization Philosophy

Key insight: We never measure bare parameters, only physical observables.

Procedure:

1. Add counterterms to the Lagrangian: $\mathcal{L} \to \mathcal{L} + \delta\mathcal{L}$
2. Choose $\delta\mathcal{L}$ to cancel infinities
3. Express results in terms of physical parameters

Example:

$$\mathcal{L} = \frac{1}{2}Z_\phi \partial_\mu \phi \partial^\mu \phi - \frac{1}{2}Z_m m^2 \phi^2 - \frac{Z_\lambda \lambda}{4!}\phi^4 \tag{11.30}$$

where Z_ϕ, Z_m, Z_λ are renormalization constants.

11.4.3 Renormalization Schemes

Different prescriptions for fixing counterterms:

On-shell scheme: Fix at physical values

- $m^2_{phys} = m^2 + \Sigma(m^2_{phys})$
- Intuitive but gauge-dependent

Minimal subtraction (MS): Remove only poles

- In dimensional regularization: $\delta\mathcal{L} \propto 1/\varepsilon$
- Gauge invariant but scale-dependent

Modified MS ($\overline{\text{MS}}$): Remove poles and universal constants

- $\delta\mathcal{L} \propto 1/\varepsilon - \gamma + \ln(4\pi)$

11.4.4 Running Couplings

Physical parameters depend on energy scale μ:

$$\mu \frac{dg}{d\mu} = \beta(g) \tag{11.31}$$

where $\beta(g)$ is the beta function.

For QED at one loop:

$$\beta(e) = \frac{e^3}{12\pi^2} \tag{11.32}$$

The running coupling:

$$\alpha(\mu) = \frac{\alpha(\mu_0)}{1 - \frac{\alpha(\mu_0)}{3\pi}\ln(\mu/\mu_0)} \tag{11.33}$$

QED coupling increases with energy (Landau pole).

11.4.5 Asymptotic Freedom and Confinement

In non-Abelian gauge theories:

$$\beta(g) = -\frac{b_0 g^3}{16\pi^2} + O(g^5) \tag{11.34}$$

where $b_0 = \frac{11N}{3} - \frac{2N_f}{3}$ for SU(N) with N_f flavors.

For QCD (N=3, $N_f < 17$): $b_0 > 0$, so:

- $g(\mu) \to 0$ as $\mu \to \infty$ (asymptotic freedom)
- $g(\mu) \to \infty$ as $\mu \to \Lambda_{QCD}$ (confinement)

This explains why quarks behave as free particles at high energy but are confined at low energy.

11.4.6 Renormalization Group Equations

The Callan-Symanzik equation describes how Green's functions change with scale:

$$\left[\mu\frac{\partial}{\partial\mu} + \beta(g)\frac{\partial}{\partial g} + n_\phi\gamma_\phi(g)\right] G^{(n)}(x_1, ..., x_n; g, \mu) = 0 \tag{11.35}$$

where γ_ϕ is the anomalous dimension.

Applications:

- Improved perturbation theory
- Connecting physics at different scales
- Critical phenomena (ε-expansion)
- Running of Standard Model parameters

11.5 Effective Action

11.5.1 Definition and Meaning

The effective action $\Gamma[\phi_c]$ is the quantum generalization of the classical action:

$$\Gamma[\phi_c] = W[J] - \int d^4x J(x)\phi_c(x) \tag{11.36}$$

where $\phi_c = \langle \phi \rangle_J = \delta W / \delta J$.

Physical interpretation:

- $\Gamma[\phi_c]$ includes all quantum corrections
- Its minimum gives the true vacuum
- $\delta\Gamma/\delta\phi_c = 0$ yields quantum equations of motion

11.5.2 Loop Expansion

The effective action has a loop expansion:

$$\Gamma[\phi_c] = S[\phi_c] + \frac{\hbar}{i}\Gamma^{(1)}[\phi_c] + \left(\frac{\hbar}{i}\right)^2 \Gamma^{(2)}[\phi_c] + ... \tag{11.37}$$

where:

- $S[\phi_c]$: Tree level (classical action)
- $\Gamma^{(1)}$: One-loop quantum corrections
- $\Gamma^{(n)}$: n-loop contributions

One-loop result:

$$\Gamma^{(1)} = \frac{1}{2}\mathrm{Tr}\ln\left[\frac{\delta^2 S}{\delta\phi\delta\phi}\right] \tag{11.38}$$

11.5.3 Effective Potential

For constant fields $\phi_c = \phi_0$:

$$V_{eff}(\phi_0) = -\frac{\Gamma[\phi_0]}{VT} \tag{11.39}$$

where V is spatial volume and T is time.

Coleman-Weinberg potential (massless ϕ^4):

$$V_{eff}(\phi) = \frac{\lambda\phi^4}{4!}\left[\ln\left(\frac{\phi^2}{\mu^2}\right) - \frac{25}{6}\right] \tag{11.40}$$

This shows:

- Radiative symmetry breaking possible
- Running of couplings encoded
- Phase transitions in field theory

11.5.4 Effective Field Theory

At low energies, integrate out heavy fields:

$$Z = \int \mathcal{D}[\Phi]\mathcal{D}[\phi]e^{iS[\Phi,\phi]/\hbar} = \int \mathcal{D}[\phi]e^{iS_{eff}[\phi]/\hbar} \tag{11.41}$$

where $S_{eff}[\phi] = -i\hbar \ln \int \mathcal{D}[\Phi]e^{iS[\Phi,\phi]/\hbar}$.

Example - Fermi theory from electroweak:

- High energy: W-boson exchange
- Low energy: Four-fermion interaction
- $G_F = \frac{g^2}{8M_W^2}$

11.5.5 Anomalous Effective Actions

For chiral fermions, the effective action is not gauge invariant:

$$\Gamma[A] = \int d^4x \left[-\frac{1}{4}F_{\mu\nu}F^{\mu\nu}\right] + \Gamma_{anom}[A] \tag{11.42}$$

where:

$$\Gamma_{anom}[A] = \frac{e^2}{48\pi^2}\int d^4x \varepsilon^{\mu\nu\rho\sigma}A_\mu\partial_\nu A_\rho\partial_\sigma A_\tau \tag{11.43}$$

This Wess-Zumino term:

- Reproduces the chiral anomaly
- Is topological (metric-independent)
- Constrains consistent gauge theories

11.5.6 Applications of Effective Actions

Symmetry breaking: Determine vacuum structure

- Higgs mechanism
- Chiral symmetry breaking in QCD

Phase transitions: Study critical behavior

- Finite temperature field theory
- Cosmological phase transitions

Solitons and instantons: Non-perturbative physics

- Monopoles, skyrmions
- Tunneling rates

Quantum corrections: Systematic calculation

- Casimir effect
- Hawking radiation

11.6 Summary

Quantum field theory elevates the principle of least action to its most powerful form. The path integral over field configurations provides a unified framework for understanding all fundamental interactions. From this action-based approach emerges the entire machinery of modern particle physics.

Key achievements of the QFT action principle:

1. **Unification of Concepts:** The same action principle governs:
    - Free and interacting fields
    - Bosons and fermions
    - Abelian and non-Abelian gauge theories
    - Massive and massless particles
2. **Symmetry Implementation:** Noether's theorem in QFT yields:
    - Conservation laws and selection rules
    - Ward identities constraining amplitudes
    - Understanding of massless particles
    - Anomalies and their physical consequences
3. **Calculational Framework:** Feynman rules from the action provide:

- Systematic perturbation theory
- Manifestly covariant calculations
- Connection to experiment via S-matrix
- Non-perturbative insights

4. **Renormalization Understanding:** The action framework clarifies:

   - Origin and treatment of infinities
   - Running of couplings with energy
   - Asymptotic freedom and confinement
   - Effective theories at different scales

5. **Effective Action Insights:** Quantum corrections included through:

   - Spontaneous symmetry breaking
   - Phase transitions
   - Vacuum structure
   - Low-energy effective theories

The principle of least action thus provides not just a formulation but a deep organizing principle for quantum field theory. It guides the construction of new theories, constrains their structure through symmetries, and provides the tools for extracting physical predictions. As we'll see in the next chapter, these same principles guide our understanding of the Standard Model and the search for physics beyond it.

Chapter 12

Particle Physics and Gauge Theories

The Standard Model of particle physics represents one of the greatest triumphs of the principle of least action. Built entirely on gauge symmetry principles, it describes all known fundamental particles and their interactions (except gravity) through a single action functional. The requirement that physics be invariant under local symmetry transformations severely constrains the possible forms of interaction, leading to a remarkably predictive theory that has been tested to extraordinary precision.

This chapter explores how gauge principles determine the structure of particle physics theories. We'll see how the simple requirement of local gauge invariance leads inevitably to the existence of force-carrying bosons, how the Higgs mechanism reconciles gauge invariance with massive particles, and how the entire Standard Model emerges from symmetry principles. We'll also explore supersymmetry, a proposed extension that further constrains the action through a symmetry between bosons and fermions. Throughout, the principle of least action serves not just as a calculational tool but as the fundamental organizing principle.

12.1 Gauge Principle and Invariant Action

12.1.1 From Global to Local Symmetry

Consider a complex scalar field with global U(1) symmetry:

$$\mathcal{L} = \partial_\mu \phi^* \partial^\mu \phi - m^2 \phi^* \phi \tag{12.1}$$

This is invariant under $\phi \to e^{i\alpha}\phi$ with constant α.

Now demand local symmetry: $\phi(x) \to e^{i\alpha(x)}\phi(x)$. The derivative term breaks this:

$$\partial_\mu \phi \to e^{i\alpha}(\partial_\mu \phi + i\partial_\mu \alpha \cdot \phi) \tag{12.2}$$

To restore invariance, we must:

1. Introduce a gauge field A_μ
2. Replace $\partial_\mu \to D_\mu = \partial_\mu + ieA_\mu$ (covariant derivative)
3. Require $A_\mu \to A_\mu - \frac{1}{e}\partial_\mu \alpha$

The invariant Lagrangian becomes:

$$\mathcal{L} = |D_\mu \phi|^2 - m^2 |\phi|^2 - \frac{1}{4} F_{\mu\nu} F^{\mu\nu} \tag{12.3}$$

where $F_{\mu\nu} = \partial_\mu A_\nu - \partial_\nu A_\mu$ ensures gauge invariant dynamics for A_μ.

12.1.2 Non-Abelian Gauge Theories

For non-Abelian groups like SU(N), the structure is richer. Consider SU(2):

$$\phi \to U(x)\phi \quad \text{where} \quad U(x) = \exp(i\alpha^a(x)\tau^a/2) \tag{12.4}$$

The gauge field becomes matrix-valued: $A_\mu = A_\mu^a \tau^a/2$

The covariant derivative:

$$D_\mu = \partial_\mu + igA_\mu \tag{12.5}$$

transforms as: $D_\mu \phi \to U(D_\mu \phi)$

The field strength tensor:

$$F_{\mu\nu} = \partial_\mu A_\nu - \partial_\nu A_\mu + ig[A_\mu, A_\nu] \tag{12.6}$$

The commutator term makes the theory non-linear—gauge bosons self-interact.

12.1.3 Yang-Mills Action

The gauge invariant action for pure Yang-Mills theory:

$$S_{YM} = -\frac{1}{4}\int d^4x \mathrm{Tr}(F_{\mu\nu}F^{\mu\nu}) \tag{12.7}$$

Expanding in components:

$$S_{YM} = -\frac{1}{4}\int d^4x F^a_{\mu\nu}F^{a\mu\nu} \tag{12.8}$$

where:

$$F^a_{\mu\nu} = \partial_\mu A^a_\nu - \partial_\nu A^a_\mu + gf^{abc}A^b_\mu A^c_\nu \tag{12.9}$$

The structure constants f^{abc} encode the Lie algebra: $[\tau^a, \tau^b] = if^{abc}\tau^c$

12.1.4 Matter Coupling

Fermions in representation R couple through:

$$\mathcal{L}_{matter} = \bar{\psi}(i\not{D} - m)\psi \tag{12.10}$$

where $\not{D} = \gamma^\mu D_\mu$ and:

$$D_\mu\psi = (\partial_\mu + igA^a_\mu T^a_R)\psi \tag{12.11}$$

T^a_R are generators in representation R.
For scalars:

$$\mathcal{L}_{scalar} = |D_\mu\phi|^2 - V(\phi) \tag{12.12}$$

The complete gauge theory:

$$\mathcal{L} = -\frac{1}{4}F^a_{\mu\nu}F^{a\mu\nu} + \bar{\psi}(i\not{D} - m)\psi + |D_\mu\phi|^2 - V(\phi) \tag{12.13}$$

12.1.5 Gauge Fixing and Ghosts

Gauge invariance means not all components of A_μ are physical. To quantize, we must fix the gauge:

- Lorenz gauge: $\partial_\mu A^\mu = 0$
- Coulomb gauge: $\nabla \cdot \mathbf{A} = 0$

- Axial gauge: $n \cdot A = 0$

In path integrals, gauge fixing introduces:

1. Gauge-fixing term: $-\frac{1}{2\xi}(\partial_\mu A^\mu)^2$
2. Faddeev-Popov ghosts: $\mathcal{L}_{ghost} = \bar{c}^a \partial_\mu (D^\mu c)^a$

Ghosts are anticommuting scalars that cancel unphysical polarizations in loops.

12.1.6 BRST Symmetry

Gauge-fixed theories possess BRST symmetry:

$$\delta A_\mu = D_\mu c \tag{12.14}$$
$$\delta c = -\frac{g}{2} f^{abc} c^b c^c \tag{12.15}$$
$$\delta \bar{c} = -\frac{1}{\xi} \partial_\mu A^\mu \tag{12.16}$$

The BRST charge Q_B satisfies $Q_B^2 = 0$.
Physical states satisfy $Q_B|\text{phys}\rangle = 0$, ensuring:

- Gauge invariance of physical results
- Unitarity of S-matrix
- Decoupling of unphysical modes

12.2 Standard Model of Particle Physics

12.2.1 The Gauge Group

The Standard Model is based on the gauge group:

$$G_{SM} = SU(3)_C \times SU(2)_L \times U(1)_Y \tag{12.17}$$

- $SU(3)_C$: Color (strong force)
- $SU(2)_L$: Weak isospin

- $U(1)_Y$: Weak hypercharge

This requires:

- 8 gluons for SU(3)
- 3 W-bosons for SU(2)
- 1 B-boson for U(1)

12.2.2 Particle Content

Quarks (3 generations):

- Left-handed doublets: $Q_L = \begin{pmatrix} u_L \\ d_L \end{pmatrix}$ transforms as $(3, 2, 1/6)$
- Right-handed singlets: u_R $(3, 1, 2/3)$, d_R $(3, 1, -1/3)$

Leptons (3 generations):

- Left-handed doublets: $L_L = \begin{pmatrix} \nu_L \\ e_L \end{pmatrix}$ transforms as $(1, 2, -1/2)$
- Right-handed singlets: e_R $(1, 1, -1)$

Higgs:

- Complex doublet: H transforms as $(1, 2, 1/2)$

Numbers indicate $(SU(3), SU(2), U(1))$ representations.

12.2.3 The Standard Model Lagrangian

$$\mathcal{L}_{SM} = \mathcal{L}_{gauge} + \mathcal{L}_{fermion} + \mathcal{L}_{Higgs} + \mathcal{L}_{Yukawa} \tag{12.18}$$

Gauge sector:

$$\mathcal{L}_{gauge} = -\frac{1}{4}G^a_{\mu\nu}G^{a\mu\nu} - \frac{1}{4}W^i_{\mu\nu}W^{i\mu\nu} - \frac{1}{4}B_{\mu\nu}B^{\mu\nu} \tag{12.19}$$

Fermion kinetic terms:

$$\mathcal{L}_{fermion} = \sum_f \bar{\psi}_f i \not{D}\psi_f \tag{12.20}$$

Higgs sector:

$$\mathcal{L}_{Higgs} = |D_\mu H|^2 - V(H) \tag{12.21}$$
$$V(H) = -\mu^2 H^\dagger H + \lambda (H^\dagger H)^2 \tag{12.22}$$

Yukawa couplings:

$$\mathcal{L}_{Yukawa} = -Y_d \bar{Q}_L H d_R - Y_u \bar{Q}_L \tilde{H} u_R - Y_e \bar{L}_L H e_R + \text{h.c.} \tag{12.23}$$

where $\tilde{H} = i\sigma_2 H^*$ is the charge conjugate Higgs.

12.2.4 Electroweak Unification

The $SU(2)_L \times U(1)_Y$ symmetry breaks to $U(1)_{EM}$ when $\langle H \rangle \neq 0$.

The gauge bosons mix:

$$\text{Photon: } A_\mu = B_\mu \cos\theta_W + W_\mu^3 \sin\theta_W \tag{12.24}$$
$$\text{Z-boson: } Z_\mu = -B_\mu \sin\theta_W + W_\mu^3 \cos\theta_W \tag{12.25}$$
$$\text{W-bosons: } W_\mu^\pm = \frac{1}{\sqrt{2}}(W_\mu^1 \mp i W_\mu^2) \tag{12.26}$$

where $\tan\theta_W = g'/g$ is the Weinberg angle.

The electromagnetic coupling: $e = g\sin\theta_W = g'\cos\theta_W$

12.2.5 QCD Sector

The $SU(3)_C$ theory of quarks and gluons:

$$\mathcal{L}_{QCD} = -\frac{1}{4} G^a_{\mu\nu} G^{a\mu\nu} + \sum_q \bar{\psi}_q (i \not{D} - m_q) \psi_q \tag{12.27}$$

Properties:

- Asymptotic freedom: $\alpha_s(\mu) \to 0$ as $\mu \to \infty$
- Confinement: Only color singlets observed
- Chiral symmetry breaking: $\langle \bar{\psi}\psi \rangle \neq 0$

Running coupling:

$$\alpha_s(\mu) = \frac{\alpha_s(\mu_0)}{1 + \frac{\alpha_s(\mu_0)}{4\pi} b_0 \ln(\mu^2/\mu_0^2)} \tag{12.28}$$

with $b_0 = 11 - 2n_f/3$.

12.2.6 Anomalies and Consistency

The Standard Model is anomaly-free—crucial for consistency:

Triangle anomaly cancellation:

$$\sum_{fermions} T_a\{T_b, T_c\} = 0 \tag{12.29}$$

For each generation:

- Color anomaly: $3 \times 2 \times (2/3 - 1/3) = 0$
- SU(2) anomaly: $3 \times (1/6) + 1 \times (-1/2) = 0$
- Mixed anomalies: All cancel

This constrains:

- Number of colors = 3
- Quarks and leptons paired
- Complete generations required

12.3 Spontaneous Symmetry Breaking

12.3.1 The Higgs Potential

The Higgs potential:

$$V(H) = -\mu^2 H^\dagger H + \lambda (H^\dagger H)^2 \tag{12.30}$$

For $\mu^2 > 0$, the minimum is at:

$$\langle H \rangle = \begin{pmatrix} 0 \\ v/\sqrt{2} \end{pmatrix} \quad \text{where} \quad v = \sqrt{\mu^2/\lambda} \approx 246 \text{ GeV} \tag{12.31}$$

This breaks $SU(2)_L \times U(1)_Y \to U(1)_{EM}$.

12.3.2 Goldstone Bosons

The Higgs field has 4 real components. After symmetry breaking:

$$H = \begin{pmatrix} 0 \\ \frac{v+h+i\chi_3}{\sqrt{2}} \end{pmatrix} \exp\left[i\frac{\chi_1\tau_1 + \chi_2\tau_2}{v}\right] \tag{12.32}$$

- h: Physical Higgs boson
- χ_1, χ_2, χ_3: Goldstone bosons

In unitary gauge, Goldstone bosons are gauged away, becoming longitudinal components of massive gauge bosons.

12.3.3 Gauge Boson Masses

From $|D_\mu H|^2$ with $\langle H \rangle \neq 0$:

$$M_W = \frac{gv}{2} \approx 80.4 \text{ GeV} \tag{12.33}$$

$$M_Z = \frac{M_W}{\cos\theta_W} \approx 91.2 \text{ GeV} \tag{12.34}$$

$$M_\gamma = 0 \tag{12.35}$$

The photon remains massless because $U(1)_{EM}$ is unbroken.
Mass eigenstates:

- $W^\pm$ get mass from eating charged Goldstones
- Z gets mass from eating neutral Goldstone
- γ remains massless (unbroken symmetry)

12.3.4 Fermion Masses

Yukawa couplings generate masses after symmetry breaking:

$$\mathcal{L}_{mass} = -\frac{v}{\sqrt{2}}(Y_d\bar{d}d + Y_u\bar{u}u + Y_e\bar{e}e) \tag{12.36}$$

Masses: $m_f = \frac{Y_f v}{\sqrt{2}}$
The mass hierarchy problem:

- $m_t \approx 173$ GeV $\Rightarrow Y_t \approx 1$
- $m_e \approx 0.5$ MeV $\Rightarrow Y_e \approx 10^{-6}$

Why such different Yukawa couplings?

12.3.5 CKM Matrix

Quark mass eigenstates differ from weak eigenstates:

$$d'_L = V_{CKM} d_L \tag{12.37}$$

The CKM matrix is unitary 3×3:

$$V_{CKM} = \begin{pmatrix} V_{ud} & V_{us} & V_{ub} \\ V_{cd} & V_{cs} & V_{cb} \\ V_{td} & V_{ts} & V_{tb} \end{pmatrix} \tag{12.38}$$

This leads to:

- Flavor-changing charged currents
- CP violation (complex phase)
- No flavor-changing neutral currents (GIM mechanism)

12.3.6 Neutrino Masses

The Standard Model has massless neutrinos, but oscillations require mass.

Possible extensions:

- Right-handed neutrinos: Dirac masses
- Majorana masses: $\nu = \nu^c$
- Seesaw mechanism: Light ν from heavy N

The PMNS matrix for leptons:

$$\nu'_L = U_{PMNS} \nu_L \tag{12.39}$$

describes neutrino mixing.

12.4 Higgs Mechanism

12.4.1 Abelian Higgs Model

Consider U(1) gauge theory with complex scalar:

$$\mathcal{L} = |D_\mu \phi|^2 - V(\phi) - \frac{1}{4} F_{\mu\nu} F^{\mu\nu} \tag{12.40}$$

where $V(\phi) = -\mu^2 |\phi|^2 + \lambda |\phi|^4$.

For $\mu^2 > 0$: $\langle \phi \rangle = \frac{v}{\sqrt{2}} e^{i\theta}$

12.4.2 Unitary Gauge

Write $\phi = \frac{v+h}{\sqrt{2}} e^{i\chi/v}$ and gauge away χ:

$$\phi \to \frac{v+h}{\sqrt{2}} \tag{12.41}$$

The Lagrangian becomes:

$$\mathcal{L} = \frac{1}{2}\partial_\mu h \partial^\mu h - \frac{1}{2}m_h^2 h^2 + \frac{1}{2}m_A^2 A_\mu A^\mu - \frac{1}{4}F_{\mu\nu}F^{\mu\nu} + \text{interactions} \tag{12.42}$$

where:

- $m_A = ev$ (gauge boson mass)
- $m_h = \sqrt{2\lambda}v$ (Higgs mass)

12.4.3 Renormalizability

The Higgs mechanism preserves renormalizability:

- Massive gauge theory naively non-renormalizable
- Spontaneous breaking maintains gauge structure
- 't Hooft proved renormalizability (1971)

Key: Gauge symmetry still present, just hidden.

12.4.4 Equivalence Theorem

At high energy $E \gg M_W$:

$$\text{Amplitude}(W_L W_L \to X) \approx \text{Amplitude}(\chi\chi \to X) \tag{12.43}$$

Longitudinal W's act like their Goldstone bosons.
This simplifies high-energy calculations:

- Use scalar diagrams instead of massive vectors
- Understand unitarity constraints
- Connect to strongly-coupled theories

12.4.5 Higgs Couplings

The Higgs couples to mass:

$$\text{Fermions: } g_{hff} = \frac{m_f}{v} \tag{12.44}$$

$$\text{W-bosons: } g_{hWW} = \frac{2M_W^2}{v} \tag{12.45}$$

$$\text{Z-boson: } g_{hZZ} = \frac{2M_Z^2}{v} \tag{12.46}$$

$$\text{Higgs self-coupling: } \lambda_{hhh} = \frac{3m_h^2}{v} \tag{12.47}$$

Discovery at LHC (2012) confirmed:

- $m_h \approx 125$ GeV
- Couplings consistent with SM
- Spin-0, even parity

12.4.6 Hierarchy Problem

Quantum corrections to Higgs mass:

$$\delta m_h^2 \sim \frac{\lambda}{16\pi^2}\Lambda^2 \tag{12.48}$$

where Λ is the cutoff scale.

For $\Lambda = M_{Planck} \sim 10^{19}$ GeV:

- Expect $m_h \sim M_{Planck}$
- Observed $m_h \sim 100$ GeV
- Requires fine-tuning to 1 part in 10^{34}

Proposed solutions:

- Supersymmetry (cancellation between bosons/fermions)
- Composite Higgs (not fundamental)
- Extra dimensions (lower Planck scale)

12.5 Supersymmetric Actions

12.5.1 Supersymmetry Algebra

Supersymmetry relates bosons and fermions:

$$Q|\text{boson}\rangle = |\text{fermion}\rangle \tag{12.49}$$
$$Q|\text{fermion}\rangle = |\text{boson}\rangle \tag{12.50}$$

The SUSY algebra:

$$\{Q_\alpha, \bar{Q}_{\dot{\beta}}\} = 2\sigma^\mu_{\alpha\dot{\beta}} P_\mu \tag{12.51}$$
$$\{Q_\alpha, Q_\beta\} = \{\bar{Q}_{\dot{\alpha}}, \bar{Q}_{\dot{\beta}}\} = 0 \tag{12.52}$$
$$[Q_\alpha, P_\mu] = 0 \tag{12.53}$$

This extends the Poincaré algebra.

12.5.2 Superfields

Supersymmetric theories are elegantly formulated using superfields:

Chiral superfield:

$$\Phi(x, \theta, \bar{\theta}) = \phi(x) + \sqrt{2}\theta\psi(x) + \theta\theta F(x) + ... \tag{12.54}$$

- ϕ: complex scalar
- ψ: Weyl fermion
- F: auxiliary field

Vector superfield:

$$V(x, \theta, \bar{\theta}) = ... + \theta\sigma^\mu\bar{\theta}A_\mu(x) + \theta\theta\bar{\theta}\bar{\lambda}(x) + ... \tag{12.55}$$

- A_μ: gauge field
- λ: gaugino
- D: auxiliary field

12.5.3 Supersymmetric Actions

The general form:

$$S = \int d^4x d^2\theta d^2\bar{\theta} K(\Phi^\dagger, \Phi, V) + \left[\int d^4x d^2\theta W(\Phi) + \text{h.c.}\right] + \int d^4x d^2\theta \tau W^\alpha W_\alpha \tag{12.56}$$

where:

- K: Kähler potential (kinetic terms)
- W: Superpotential (interactions)
- W^α: Field strength superfield

12.5.4 The MSSM

The Minimal Supersymmetric Standard Model doubles the particle content:

Chiral supermultiplets:

- Quarks → Squarks ($\tilde{q}_L, \tilde{q}_R$)
- Leptons → Sleptons ($\tilde{l}_L, \tilde{l}_R$)
- Higgs → Higgsinos ($\tilde{H}_u, \tilde{H}_d$)

Vector supermultiplets:

- Gluons → Gluinos ($\tilde{g}$)
- W,B → Winos, Bino ($\tilde{W}, \tilde{B}$)

The MSSM Lagrangian preserves gauge invariance and adds soft SUSY breaking terms.

12.5.5 SUSY Breaking

Supersymmetry must be broken (no observed selectrons at m_e).

Soft breaking terms:

$$\mathcal{L}_{soft} = -m_\phi^2|\phi|^2 - (A\lambda\phi^3 + B\mu\phi_1\phi_2 + \text{h.c.}) - \frac{1}{2}M\lambda\lambda \tag{12.57}$$

These preserve renormalizability.

Breaking mechanisms:

- Gravity mediation: $M_{soft} \sim F/M_{Planck}$
- Gauge mediation: $M_{soft} \sim \frac{\alpha}{4\pi} F/M_{messenger}$
- Anomaly mediation: $M_{soft} \sim \frac{\alpha}{4\pi} F/M_{Planck}$

12.5.6 Phenomenological Consequences

SUSY solves several problems:

Hierarchy problem: Cancellation between bosons/fermions

$$\delta m_h^2|_{fermion} = -\frac{\lambda_f^2}{16\pi^2}\Lambda^2 \tag{12.58}$$

$$\delta m_h^2|_{scalar} = +\frac{\lambda_s^2}{16\pi^2}\Lambda^2 \tag{12.59}$$

Natural if $\lambda_f \approx \lambda_s$ (SUSY relates them)

Gauge coupling unification: RG evolution changes

- $\alpha_1, \alpha_2, \alpha_3$ meet at $M_{GUT} \sim 10^{16}$ GeV

Dark matter: Lightest supersymmetric particle (LSP)

- R-parity conservation $\Rightarrow$ LSP stable
- Neutralino dark matter candidate

Constraints: No SUSY particles found yet

- $m_{\tilde{g}} > 2$ TeV
- $m_{\tilde{q}} > 1.5$ TeV
- Natural SUSY increasingly constrained

12.6 Summary

The principle of least action reaches its full power in gauge theories and particle physics. Starting from the simple requirement of local gauge invariance, we're led inexorably to:

1. **Force-Carrying Bosons:** Local symmetry requires gauge fields that mediate interactions. The action principle determines their dynamics and self-interactions.

2. **Constrained Interactions:** Gauge invariance severely restricts possible interaction terms. This explains why nature uses only certain forces with specific properties.
3. **Unification:** Different forces emerge as aspects of a single gauge theory. The electroweak theory unifies electromagnetic and weak forces; grand unification extends this idea.
4. **Mass Generation:** The Higgs mechanism reconciles gauge invariance with massive particles, preserving the action's gauge structure while breaking the symmetry spontaneously.
5. **Predictive Power:** The Standard Model action makes precise predictions:
    - Existence and properties of W, Z bosons (confirmed)
    - Higgs boson mass range (confirmed at 125 GeV)
    - Anomaly cancellation constraining particle content
    - Running couplings and asymptotic freedom
6. **Beyond the Standard Model:** Extensions maintain the action principle:
    - Supersymmetry: Relates bosons and fermions
    - Grand unification: Larger gauge groups
    - Extra dimensions: Gauge theory in higher dimensions

The gauge principle demonstrates that symmetry determines dynamics. Rather than postulating forces, we identify symmetries and let the requirement of local invariance dictate the interactions. This approach has been spectacularly successful, describing all non-gravitational physics in a unified framework.

Yet puzzles remain: the hierarchy problem, dark matter, neutrino masses, and the strong CP problem all suggest physics beyond the Standard Model. The principle of least action, having guided us this far, continues to illuminate the path forward. Whether through supersymmetry, extra dimensions, or yet-unknown principles, future theories will build on the action foundation that has proven so powerful in revealing nature's fundamental laws.

Chapter 13

Optics and Fermat's Principle

Optics, where the principle of least action began with Fermat's insight about light paths, continues to be a fertile ground for variational methods. Modern optics extends far beyond ray tracing and linear wave propagation to encompass nonlinear phenomena, artificially structured materials with exotic properties, and quantum effects where single photons exhibit both wave and particle characteristics. The principle of least action provides a unified framework for understanding these diverse phenomena.

This chapter explores how Fermat's principle generalizes to modern optical systems. We'll see how nonlinear effects modify the action, leading to phenomena like solitons and frequency conversion. We'll examine metamaterials—engineered structures that bend light in ways impossible with natural materials—through the lens of effective actions. Finally, we'll explore quantum optics, where the classical action acquires quantum corrections and new variational principles emerge. Throughout, the connection between optics and mechanics, first noted by Hamilton, continues to provide deep insights.

13.1 Revisiting Fermat's Principle

13.1.1 Classical Formulation and Modern Context

Fermat's principle states that light travels between two points along the path of stationary optical length:

$$\delta \int n(s)ds = 0 \tag{13.1}$$

where $n(s)$ is the refractive index along the path and s is arc length.

In modern notation, the optical path length is:

$$S = \int_A^B n(\mathbf{r})ds = \frac{1}{c}\int_A^B n(\mathbf{r})\left|\frac{d\mathbf{r}}{dt}\right| dt \tag{13.2}$$

This is formally identical to the action for a particle in a potential, with:

- "Lagrangian": $L = n(\mathbf{r})|\dot{\mathbf{r}}|$
- "Energy": E = constant (Fermat's principle assumes monochromatic light)
- "Potential": $V(\mathbf{r}) = E - n(\mathbf{r})$

13.1.2 From Rays to Waves

The connection between ray optics and wave optics parallels that between classical and quantum mechanics:

Wave equation: $\nabla^2\psi + k_0^2 n^2(\mathbf{r})\psi = 0$

Assuming $\psi = A(\mathbf{r})e^{ik_0 S(\mathbf{r})}$ with slowly varying amplitude:

$$|\nabla S|^2 = n^2(\mathbf{r}) \quad \text{(eikonal equation)} \tag{13.3}$$

This is the optical Hamilton-Jacobi equation, with S playing the role of Hamilton's principal function.

13.1.3 Lagrangian and Hamiltonian Formulations

The ray Lagrangian:

$$L = n(\mathbf{r})\sqrt{\dot{x}^2 + \dot{y}^2 + \dot{z}^2} \tag{13.4}$$

Introducing the ray parameter τ:

$$L = n(\mathbf{r})\sqrt{g_{ij}\frac{dx^i}{d\tau}\frac{dx^j}{d\tau}} \tag{13.5}$$

The canonical momenta:

$$p_i = \frac{\partial L}{\partial \dot{x}^i} = \frac{n(\mathbf{r})\dot{x}^i}{|\dot{\mathbf{r}}|} \tag{13.6}$$

The Hamiltonian (for rays parametrized by z):

$$H = -p_z = -n(\mathbf{r})\sqrt{1 - \frac{p_x^2}{n^2} - \frac{p_y^2}{n^2}} \tag{13.7}$$

13.1.4 Gradient-Index Optics

In gradient-index (GRIN) materials, $n = n(\mathbf{r})$ varies continuously. The ray equations become:

$$\frac{d}{ds}\left(n\frac{d\mathbf{r}}{ds}\right) = \nabla n \tag{13.8}$$

For a parabolic profile $n(r) = n_0(1 - \alpha^2 r^2/2)$:

- Rays follow sinusoidal paths
- Self-focusing with period $2\pi/\alpha$
- Basis for GRIN lenses and fibers

The action approach naturally handles:

- Arbitrary index profiles
- Ray invariants and conservation laws
- Connection to quantum corrections

13.1.5 Anisotropic Media

In anisotropic crystals, the index depends on propagation direction:

$$n = n(\mathbf{r}, \hat{\mathbf{k}}) \tag{13.9}$$

The generalized Fermat principle:

$$\delta \int n(\mathbf{r}, \hat{\mathbf{t}}) ds = 0 \tag{13.10}$$

where $\hat{\mathbf{t}}$ is the unit tangent.

This leads to:

- Ordinary and extraordinary rays
- Double refraction
- Conical refraction

The Lagrangian formulation:

$$L = \sqrt{g_{ij}(\mathbf{r}, \dot{\mathbf{r}})\dot{x}^i \dot{x}^j} \tag{13.11}$$

where g_{ij} is the optical metric tensor.

13.1.6 Modern Applications

Fermat's principle continues to guide optical design:

Transformation optics: Design materials to guide light along prescribed paths

- Invisibility cloaks
- Perfect lenses
- Light harvesting

Optimization: Find index profiles for specific functions

- Minimize aberrations
- Maximize light collection

- Design photonic crystals

Computational methods: Ray tracing from action principles

- Symplectic integrators preserve ray invariants
- Action-based mesh refinement
- Global optimization using action landscapes

13.2 Nonlinear Optics

13.2.1 Nonlinear Polarization

In intense fields, the material polarization becomes nonlinear:

$$\mathbf{P} = \varepsilon_0(\chi^{(1)}\mathbf{E} + \chi^{(2)}\mathbf{E}^2 + \chi^{(3)}\mathbf{E}^3 + ...) \tag{13.12}$$

The action for nonlinear optics:

$$S = \int d^4x \left[\frac{1}{2}\varepsilon_0 E^2 - \frac{1}{2}\mu_0 H^2 - \mathbf{P} \cdot \mathbf{E}\right] \tag{13.13}$$

Expanding $\mathbf{P}[\mathbf{E}]$:

$$S = S_{linear} + \int d^4x \left[-\frac{\varepsilon_0\chi^{(2)}E^3}{3} - \frac{\varepsilon_0\chi^{(3)}E^4}{4} + ...\right] \tag{13.14}$$

13.2.2 Second-Order Nonlinearity

For $\chi^{(2)} \neq 0$ (non-centrosymmetric crystals):
The interaction Lagrangian:

$$\mathcal{L}_{int} = -\varepsilon_0\chi^{(2)}E_1E_2E_3 \tag{13.15}$$

This enables:

- Second harmonic generation: $\omega + \omega \to 2\omega$
- Sum frequency generation: $\omega_1 + \omega_2 \to \omega_3$
- Parametric down-conversion: $\omega \to \omega_1 + \omega_2$

Phase matching condition from stationary phase:

$$\mathbf{k}_3 = \mathbf{k}_1 + \mathbf{k}_2 \tag{13.16}$$

13.2.3 Self-Focusing and Solitons

For $\chi^{(3)} > 0$ (Kerr nonlinearity), $n = n_0 + n_2 I$:

The nonlinear Schrödinger equation:

$$i\frac{\partial A}{\partial z} + \frac{1}{2k}\nabla_\perp^2 A + \frac{k_0 n_2}{n_0}|A|^2 A = 0 \tag{13.17}$$

This derives from the Lagrangian:

$$\mathcal{L} = \frac{i}{2}\left(A^*\frac{\partial A}{\partial z} - A\frac{\partial A^*}{\partial z}\right) - \frac{|\nabla_\perp A|^2}{2k} + \frac{k_0 n_2}{2n_0}|A|^4 \tag{13.18}$$

Soliton solutions balance diffraction and nonlinearity:

$$A(r, z) = A_0 \text{sech}(r/w) \exp(i\beta z) \tag{13.19}$$

13.2.4 Optical Parametric Oscillators

The action for parametric interaction:

$$S = \int d^4x \left[\frac{1}{2}\sum_i (E_i^2 - c^2 B_i^2) - gE_p E_s E_i\right] \tag{13.20}$$

where p = pump, s = signal, i = idler.

- Below threshold: Classical equations give exponential gain
- Above threshold: Steady-state oscillation

The quantum action includes:

$$S_{quantum} = S_{classical} + \hbar \int d^4x \sum_i a_i^* \left(i\frac{\partial}{\partial t} - \omega_i\right) a_i \tag{13.21}$$

Leading to squeezed states and entanglement.

13.2.5 Variational Methods in Nonlinear Optics

The variational approach provides approximate soliton solutions:
Trial function: $A(r, z) = f(r/w(z)) \exp[i\phi(r, z)]$
Effective Lagrangian:

$$L_{eff} = \int_0^\infty \mathcal{L}[A_{trial}] 2\pi r dr \tag{13.22}$$

Varying with respect to $w(z)$, $\phi(z)$ gives evolution equations.
Applications:

- Spatial soliton propagation
- Pulse compression
- Mode-locking dynamics

13.2.6 Extreme Nonlinear Optics

At ultra-high intensities, perturbation theory fails:

High harmonic generation: Strong field ionization + recombination

$$S = \int dt \left[\frac{1}{2}m\dot{x}^2 + eE(t)x - I_p\right] \tag{13.23}$$

Saddle point analysis yields:

- Cutoff energy: $3.17U_p + I_p$
- Attosecond pulse generation

Relativistic nonlinearities: $v \sim c$ modifies the action

- $\gamma = 1/\sqrt{1 - v^2/c^2}$ becomes important

New phenomena:

- Relativistic self-focusing
- Wakefield acceleration
- Radiation reaction effects

13.3 Metamaterials

13.3.1 Effective Medium Theory

Metamaterials are structured materials with subwavelength features. The effective action:

$$S_{eff} = \int d^4x \left[\frac{1}{2}\varepsilon_{eff}(\omega)E^2 - \frac{1}{2}\mu_{eff}(\omega)H^2\right] \tag{13.24}$$

where ε_{eff}, μ_{eff} are derived from the microscopic structure.

For split-ring resonators + wires:

$$\varepsilon_{eff}(\omega) = 1 - \frac{\omega_p^2}{\omega^2 - \omega_e^2} \tag{13.25}$$

$$\mu_{eff}(\omega) = 1 - \frac{F\omega^2}{\omega^2 - \omega_m^2} \tag{13.26}$$

Negative index: $\varepsilon < 0$ and $\mu < 0$ simultaneously.

13.3.2 Transformation Optics

The equivalence between coordinate transformations and material properties:

Under $\mathbf{x}' = \mathbf{f}(\mathbf{x})$:

$$\varepsilon'_{ij} = (\det J)^{-1} J_{ik} J_{jl} \varepsilon_{kl} \tag{13.27}$$

$$\mu'_{ij} = (\det J)^{-1} J_{ik} J_{jl} \mu_{kl} \tag{13.28}$$

where $J_{ij} = \partial x'_i / \partial x_j$.

This enables:

- Invisibility cloaks: $r' = R_1 + r(R_2 - R_1)/R_2$
- Perfect lenses: Pendry's superlens
- Illusion optics: Make objects appear different

13.3.3 Photonic Crystals

Periodic structures with photonic band gaps. The action:

$$S = \int d^4x \left[\frac{1}{2}\varepsilon(\mathbf{r})E^2 - \frac{1}{2}\mu_0 H^2\right] \tag{13.29}$$

where $\varepsilon(\mathbf{r}+\mathbf{R}) = \varepsilon(\mathbf{r})$ for lattice vectors $\mathbf{R}$.

Bloch's theorem: $\mathbf{E}(\mathbf{r}) = \mathbf{u_k}(\mathbf{r})e^{i\mathbf{k}\cdot\mathbf{r}}$

The variational principle for band structure:

$$\delta \int_{cell} \left[\frac{|\nabla \times \mathbf{u_k}|^2}{\varepsilon(\mathbf{r})} - \omega^2|\mathbf{u_k}|^2\right] d^3r = 0 \tag{13.30}$$

Applications:

- Photonic band gap materials
- Slow light
- Optical transistors

13.3.4 Hyperbolic Metamaterials

Materials with indefinite permittivity tensor:

$$\varepsilon_x = \varepsilon_y > 0, \quad \varepsilon_z < 0 \tag{13.31}$$

The dispersion relation:

$$\frac{k_x^2}{|\varepsilon_z|} + \frac{k_z^2}{\varepsilon_x} = \frac{\omega^2}{c^2} \tag{13.32}$$

Hyperbolic isofrequency surfaces enable:

- Super-resolution imaging
- Enhanced spontaneous emission
- Thermal hyperconductivity

The action formulation naturally handles:

- Spatial dispersion
- Nonlocal effects
- Loss and gain

13.3.5 Active and Nonlinear Metamaterials

Incorporating gain and nonlinearity:

$$S = \int d^4x \left[\frac{1}{2}(\varepsilon' - i\varepsilon'')E^2 - \frac{1}{2}\mu_0 H^2 + \mathcal{L}_{NL} \right] \tag{13.33}$$

where $\varepsilon'' < 0$ represents gain.

PT-symmetric systems: Balance gain and loss

- Exceptional points
- Unidirectional invisibility
- Coherent perfect absorption

Nonlinear metamaterials:

- Bistability and switching
- Soliton formation
- Frequency conversion

13.3.6 Topological Photonics

Metamaterials with topological properties:

The Berry phase for photonic bands:

$$\gamma_n = i \oint_C \langle u_{n\mathbf{k}} | \nabla_{\mathbf{k}} | u_{n\mathbf{k}} \rangle \cdot d\mathbf{k} \tag{13.34}$$

Chern number:

$$C = \frac{1}{2\pi} \int_{BZ} (\nabla_{\mathbf{k}} \times \mathbf{A}_n) \cdot d\mathbf{S} \tag{13.35}$$

where $\mathbf{A}_n = i\langle u_{n\mathbf{k}} | \nabla_{\mathbf{k}} | u_{n\mathbf{k}} \rangle$.

Protected edge states from bulk topology:

- Unidirectional propagation
- Robust against disorder
- Topological lasers

13.4 Variational Principles in Quantum Optics

13.4.1 Quantum Electromagnetic Action

The action for quantum electrodynamics in matter:

$$S = \int d^4x \left[-\frac{1}{4} F_{\mu\nu} F^{\mu\nu} + \bar{\psi}(i \not{D} - m)\psi + \mathcal{L}_{matter} \right] \tag{13.36}$$

In the Coulomb gauge ($\nabla \cdot \mathbf{A} = 0$):

$$H = \int d^3x \left[\frac{1}{2}(E^2 + B^2) + \psi^\dagger(-i\boldsymbol{\alpha} \cdot \nabla + \beta m)\psi + H_{int} \right] \tag{13.37}$$

Quantization: $[A_i(\mathbf{x}), \Pi_j(\mathbf{y})] = i\hbar\delta_{ij}^{\perp}(\mathbf{x} - \mathbf{y})$

13.4.2 Coherent State Path Integrals

For the quantized field, use coherent states $|\alpha\rangle$:

$$K = \int \mathcal{D}[\alpha^*, \alpha] \exp\left\{ \frac{i}{\hbar} \int dt[i\hbar\alpha^*\dot{\alpha} - H(\alpha^*, \alpha)] \right\} \tag{13.38}$$

where $H(\alpha^*, \alpha) = \langle\alpha|\hat{H}|\alpha\rangle$.

For a mode with frequency ω:

$$S = \int dt[i\hbar\alpha^*\dot{\alpha} - \hbar\omega|\alpha|^2] \tag{13.39}$$

This reproduces:

- Classical equations for $\langle\alpha\rangle$
- Quantum fluctuations around classical paths
- Vacuum energy $\hbar\omega/2$

13.4.3 Jaynes-Cummings Model

Atom-field interaction in a cavity:

$$H = \hbar\omega a^\dagger a + \frac{\hbar\omega_0}{2}\sigma_z + \hbar g(a^\dagger\sigma_- + a\sigma_+) \tag{13.40}$$

The action:

$$S = \int dt[i\hbar\langle\psi|\partial_t|\psi\rangle - \langle\psi|H|\psi\rangle] \tag{13.41}$$

Variational ansatz:

$$|\psi\rangle = c_{g,n}|g, n\rangle + c_{e,n-1}|e, n-1\rangle \tag{13.42}$$

Leads to Rabi oscillations:

- Vacuum Rabi frequency: $\Omega = 2g\sqrt{n}$
- Collapse and revival
- Entanglement generation

13.4.4 Quantum Fluctuations and Casimir Effect

Zero-point fluctuations modify the electromagnetic action:

$$E_{Casimir} = \frac{\hbar c}{2}\sum_{\mathbf{k}}\sum_{pol}\omega_{\mathbf{k}} \tag{13.43}$$

For parallel plates separated by d:

$$E(d) - E(\infty) = -\frac{\pi^2\hbar c}{720d^3}A \tag{13.44}$$

The force per unit area:

$$F = -\frac{\pi^2\hbar c}{240d^4} \tag{13.45}$$

Variational calculation using trial fields satisfying boundary conditions.

13.4.5 Parametric Down-Conversion

The interaction Hamiltonian:

$$H_{int} = \hbar\chi^{(2)} \int d^3x E_p(a_s^\dagger a_i^\dagger + a_s a_i) \tag{13.46}$$

where E_p is the classical pump field.

The two-mode squeezed state:

$$|\psi\rangle = \exp\left[r(a_s^\dagger a_i^\dagger - a_s a_i)\right]|0\rangle \tag{13.47}$$

Variational parameter r determined by:

$$\frac{\partial\langle H\rangle}{\partial r} = 0 \tag{13.48}$$

Properties:

- EPR correlations
- Sub-Poissonian statistics
- Quantum key distribution

13.4.6 Open Quantum Systems

Master equation for a cavity mode with loss:

$$\frac{\partial\rho}{\partial t} = -\frac{i}{\hbar}[H,\rho] + \kappa(2a\rho a^\dagger - a^\dagger a\rho - \rho a^\dagger a) \tag{13.49}$$

Variational principle using the Liouville-von Neumann action:

$$S = \int dt\mathrm{Tr}\left\{\rho\left[i\partial_t - \frac{1}{\hbar}H\right]\right\} + \text{dissipative terms} \tag{13.50}$$

Applications:

- Laser theory
- Quantum feedback control
- Decoherence in quantum optics

13.5 Summary

The principle of least action in optics has evolved far beyond Fermat's original insight about light rays. Modern applications span an enormous range:

1. **Generalized Ray Optics:** Fermat's principle extends to:
    - Gradient-index media with continuous variation
    - Anisotropic crystals with direction-dependent propagation
    - Transformation optics designing materials for specific ray paths
2. **Nonlinear Phenomena:** The action formulation naturally incorporates:
    - Frequency conversion processes through interaction terms
    - Soliton formation balancing nonlinearity and dispersion
    - Extreme nonlinear optics in ultra-intense fields
3. **Metamaterials:** Engineered structures exhibit:
    - Negative refraction from designed effective parameters
    - Cloaking and illusion from transformation optics
    - Topological protection from geometric phases
4. **Quantum Optics:** Quantum corrections to the classical action yield:
    - Vacuum fluctuations and the Casimir effect
    - Squeezed states and entanglement
    - Open system dynamics and decoherence

The unifying theme is that light, whether treated as rays, waves, or photons, follows paths determined by extremizing an appropriate action. This variational perspective:

- Reveals deep connections between optics and mechanics
- Provides systematic approximation methods

- Guides the design of novel optical systems
- Extends naturally from classical to quantum regimes

As optical technology advances—from integrated photonics to quantum information processing—the principle of least action continues to provide both fundamental understanding and practical design tools. The same variational principles that Fermat used to explain refraction now guide the development of metamaterial cloaks, quantum light sources, and topological photonic devices. This enduring relevance demonstrates the profound nature of the action principle, transcending its historical origins to remain at the forefront of optical science.

Chapter 14

Complex Systems

Complex systems—from turbulent fluids to living organisms, from economic markets to neural networks—challenge our traditional understanding of physics. These systems exhibit emergent behavior, self-organization, and adaptation that seem to transcend the simple mechanical principles from which they arise. Remarkably, the principle of least action extends into this realm, providing insights into how order emerges from chaos and how systems far from equilibrium organize themselves.

This chapter explores how variational principles illuminate complex systems. We'll see how thermodynamics itself can be formulated as an action principle, how Onsager's variational approach describes irreversible processes, and how dissipative structures spontaneously organize to minimize entropy production. From Prigogine's far-from-equilibrium thermodynamics to modern theories of self-organization, the action principle continues to reveal fundamental organizing principles in complexity.

14.1 Principle of Least Action in Thermodynamics

14.1.1 The Thermodynamic Action

While thermodynamics traditionally focuses on equilibrium states, a dynamical formulation exists based on action principles. For a system with thermodynamic coordinates $\{X_i\}$ (volume, magnetization, etc.)

and conjugate forces $\{Y_i\}$ (pressure, magnetic field, etc.), we can define:

$$S_{thermo} = \int_{t_1}^{t_2} L_{thermo} dt \tag{14.1}$$

where the thermodynamic Lagrangian is:

$$L_{thermo} = \sum_i Y_i \dot{X}_i - \Phi(X_i, Y_i) \tag{14.2}$$

Here Φ is a thermodynamic potential (internal energy, enthalpy, etc.).

This formulation reveals thermodynamics as a contact geometry problem, with the first law as a constraint:

$$dU = TdS - PdV + \mu dN + ... \tag{14.3}$$

14.1.2 Maximum Entropy Principle

The equilibrium state maximizes entropy subject to constraints. This variational principle:

$$\delta \left[S - \sum_i \lambda_i(\langle E_i \rangle - E_i^0) \right] = 0 \tag{14.4}$$

yields the canonical distribution:

$$\rho = \frac{1}{Z} \exp\left(-\beta \sum_i \lambda_i E_i \right) \tag{14.5}$$

where $\beta = 1/kT$ emerges as a Lagrange multiplier.

The action formulation:

$$S_{info} = -\int dt \sum_i p_i \ln\left(\frac{p_i}{p_i^0}\right) \tag{14.6}$$

measures information change from prior p_i^0 to posterior p_i.

14.1.3 Fluctuation Theorems

Near equilibrium, fluctuations satisfy Onsager's regression hypothesis. The probability of a fluctuation path $\{X(t)\}$ is:

$$P[X(t)] \propto \exp\left(-\frac{S_{fluct}[X(t)]}{kT}\right) \tag{14.7}$$

where the fluctuation action is:

$$S_{fluct} = \frac{1}{2}\int dt \sum_{ij}(\dot{X}_i - v_i)M_{ij}(\dot{X}_j - v_j) \tag{14.8}$$

Here v_i are mean velocities and M_{ij} is the mobility matrix. This leads to:

- Fluctuation-dissipation theorem
- Jarzynski equality: $\langle \exp(-W/kT)\rangle = \exp(-\Delta F/kT)$
- Crooks fluctuation theorem

14.1.4 Thermodynamic Geometry

The space of equilibrium states has a natural Riemannian geometry. The metric:

$$g_{ij} = -\frac{\partial^2 S}{\partial E_i \partial E_j} \tag{14.9}$$

measures thermodynamic distance between states.

The geodesic equation:

$$\frac{d^2X^i}{d\tau^2} + \Gamma^i_{jk}\frac{dX^j}{d\tau}\frac{dX^k}{d\tau} = 0 \tag{14.10}$$

describes optimal thermodynamic processes.

Applications:

- Finite-time thermodynamics
- Efficiency at maximum power
- Thermodynamic length and dissipation

14.1.5 Quantum Thermodynamics

For quantum systems, the von Neumann entropy:

$$S = -k\text{Tr}(\rho \ln \rho) \tag{14.11}$$

The quantum thermodynamic action:

$$S_{quantum} = \int dt[\text{Tr}(\dot{\rho}U) - \beta\langle H\rangle + \text{work terms}] \tag{14.12}$$

This yields:

- Quantum Jarzynski equality
- Information-thermodynamics connection
- Maxwell's demon resolution

14.1.6 Critical Phenomena

Near critical points, the free energy has scaling form:

$$F(t, h) = |t|^{2-\alpha} f(h/|t|^{\Delta}) \tag{14.13}$$

where $t = (T - T_c)/T_c$.

The renormalization group is a variational principle:

$$\frac{\partial S_{eff}}{\partial \Lambda} = 0 \tag{14.14}$$

where Λ is the cutoff scale.

This determines:

- Critical exponents
- Universality classes
- Crossover behavior

14.2 Non-equilibrium Thermodynamics

14.2.1 Onsager's Variational Principle

For systems near equilibrium with fluxes J_i and forces X_i:

$$J_i = \sum_j L_{ij} X_j \tag{14.15}$$

Onsager showed the phenomenological coefficients satisfy:

$$L_{ij} = L_{ji} \quad \text{(reciprocity)} \tag{14.16}$$

This follows from the variational principle:

$$\delta \int dt [\dot{\Phi} - \Psi] = 0 \tag{14.17}$$

where:

- $\dot{\Phi}$ = entropy production rate
- $\Psi = \frac{1}{2} \sum_{ij} R_{ij} J_i J_j$ (dissipation function)
- $R_{ij} = (L^{-1})_{ij}$

14.2.2 Minimum Entropy Production

For steady states with fixed boundary conditions:

$$\delta \dot{S} = \delta \int dV \sigma(\mathbf{r}) = 0 \tag{14.18}$$

where σ is the local entropy production density:

$$\sigma = \sum_i \mathbf{J}_i \cdot \mathbf{X}_i \tag{14.19}$$

This principle determines:

- Steady-state profiles
- Coupled transport coefficients
- Thermoelectric effects

Example - Heat conduction:

$$\sigma = \mathbf{J}_q \cdot \nabla(1/T) \tag{14.20}$$

Minimizing with $\mathbf{J}_q = -\kappa \nabla T$ gives Fourier's law.

14.2.3 Maximum Entropy Production

Far from equilibrium, some systems maximize entropy production:

$$\delta \int \dot{S} dt = 0 \quad \text{(maximum)} \tag{14.21}$$

This controversial principle applies to:

- Turbulent transport
- Planetary atmospheres
- Biological systems

The competition between min/max principles leads to:

- Pattern formation
- Self-organized criticality
- Edge of chaos dynamics

14.2.4 Extended Irreversible Thermodynamics

Including memory effects and spatial correlations:

$$S = S_{eq} - \frac{1}{2} \int dt' \int dV \sum_{ij} \alpha_{ij}(t - t') J_i(t) J_j(t') \tag{14.22}$$

The generalized evolution equations:

$$\tau_{ij} \frac{\partial J_i}{\partial t} + J_i = L_{ij} X_j + \text{higher order terms} \tag{14.23}$$

This describes:

- Hyperbolic heat conduction
- Viscoelastic effects
- Non-local transport

14.2.5 Mesoscopic Non-equilibrium Thermodynamics

Between microscopic and macroscopic scales:

$$S_{meso} = \int dt \left[\sum_i p_i \dot{X}_i - k \sum_i p_i \ln p_i \right] \tag{14.24}$$

where p_i is the probability of mesoscopic state i.

Master equation from variational principle:

$$\frac{\partial p_i}{\partial t} = \sum_j (W_{ij} p_j - W_{ji} p_i) \tag{14.25}$$

Applications:

- Single molecule experiments
- Molecular motors
- Stochastic thermodynamics

14.2.6 Information Thermodynamics

Information processing has thermodynamic costs:

$$S_{info} = k[H(X) - H(X|Y)] \tag{14.26}$$

where H is Shannon entropy.

Landauer's principle: Erasing one bit costs:

$$W \geq kT \ln 2 \tag{14.27}$$

The action for measurement and feedback:

$$S = S_{system} + S_{measurement} + S_{correlation} \tag{14.28}$$

This explains:

- Szilard engine operation
- Quantum measurement backaction
- Thermodynamic computing limits

14.3 Dissipative Structures

14.3.1 Far-from-Equilibrium Systems

When driven far from equilibrium, systems can spontaneously organize:

$$\frac{\partial u}{\partial t} = F(u, \nabla u, \nabla^2 u, \lambda) \tag{14.29}$$

where λ measures distance from equilibrium.

Linear stability analysis:

$$u = u_0 + \varepsilon e^{\sigma t + i\mathbf{k}\cdot\mathbf{r}} \tag{14.30}$$

yields dispersion relation $\sigma(k, \lambda)$.

For $\lambda > \lambda_c$: $\sigma > 0$ for some $k \Rightarrow$ instability $\Rightarrow$ pattern formation

14.3.2 Reaction-Diffusion Systems

The paradigmatic example:

$$\frac{\partial u}{\partial t} = D_u \nabla^2 u + f(u, v) \tag{14.31}$$

$$\frac{\partial v}{\partial t} = D_v \nabla^2 v + g(u, v) \tag{14.32}$$

Variational formulation exists for gradient systems:

$$\frac{\partial u_i}{\partial t} = -M_{ij} \frac{\delta F}{\delta u_j} \tag{14.33}$$

where F is a Lyapunov functional:

$$F = \int dV [\ell(u, \nabla u) + V(u)] \tag{14.34}$$

Examples:

- Turing patterns
- Chemical waves
- Morphogenesis

14.3.3 Convection and Turbulence

Rayleigh-Bénard convection emerges from competing effects:
Buoyancy action: $S_{buoy} = \int dtdV \rho g\alpha T v_z$
Dissipation: $S_{diss} = -\int dtdV[\nu(\nabla v)^2 + \kappa(\nabla T)^2]$
The Rayleigh number:

$$Ra = \frac{g\alpha\Delta T L^3}{\nu\kappa} \tag{14.35}$$

For $Ra > Ra_c \approx 1708$: Convection onset
Variational principles for turbulence:

- Maximum entropy production
- Minimum enstrophy
- Maximum mixing

14.3.4 Biological Dissipative Structures

Living systems maintain organization through energy flux:

$$\frac{dS}{dt} = \frac{dS_i}{dt} + \frac{dS_e}{dt} \tag{14.36}$$

where:

- $dS_i/dt > 0$ (internal production)
- $dS_e/dt < 0$ (export to environment)
- $dS/dt < 0$ (decreasing entropy)

Examples:

- Metabolic cycles
- Circadian rhythms
- Morphogenetic fields

The action for growth:

$$S_{growth} = \int dt[R(\mu) - D(X)] \tag{14.37}$$

where R is reproduction rate and D is death rate.

14.3.5 Spatiotemporal Chaos

Between order and randomness:

$$\frac{\partial u}{\partial t} = \nabla^2 u + u - u^3 + f(\mathbf{r}, t) \tag{14.38}$$

Shows:

- Strange attractors
- Intermittency
- Spatiotemporal complexity

Variational characterization:

- Minimize prediction error
- Maximize Kolmogorov-Sinai entropy
- Unstable periodic orbit expansion

14.3.6 Active Matter

Self-propelled particles with energy input:

$$\dot{\mathbf{r}}_i = v_0 \hat{\mathbf{n}}_i + \sum_j \mathbf{F}_{ij} \tag{14.39}$$

$$\dot{\theta}_i = \xi_i + \sum_j G_{ij} \tag{14.40}$$

Collective phenomena:

- Flocking (Vicsek model)
- Active turbulence
- Topological defects

Variational principles:

- Minimum dissipation rate
- Maximum polar order
- Optimal information transfer

14.4 Self-organization and Principle of Least Action

14.4.1 Emergence and Downward Causation

Complex systems show emergent properties not present in components:
Microscopic action: $S_{micro} = \sum_i S_i$
Macroscopic action: $S_{macro} = S_{collective}$
The relationship:

$$S_{macro} = -kT \ln Z_{micro} + \text{constraints} \tag{14.41}$$

Emergent variables X satisfy:

$$\frac{\delta S_{macro}}{\delta X} = \left\langle \frac{\delta S_{micro}}{\delta x} \right\rangle \Bigg|_X \tag{14.42}$$

This "downward causation" constrains microscopic dynamics.

14.4.2 Synergetics and Order Parameters

Haken's synergetics identifies order parameters ξ that govern dynamics:
Fast variables: $\dot{x} = f(x, \xi)$
Slow variables: $\dot{\xi} = g(\xi, \langle x \rangle)$
The slaving principle:

$$x = h(\xi) + \text{fluctuations} \tag{14.43}$$

Potential for order parameters:

$$V(\xi) = -\int^{\xi} g(\xi', 0) d\xi' \tag{14.44}$$

Near transitions:

$$V(\xi) = a\xi^2 + b\xi^4 + ... \tag{14.45}$$

14.4.3 Self-organized Criticality

Systems evolve to critical states without tuning:
Sandpile model:

- Add sand: $h_i \to h_i + 1$
- Topple if $h_i > h_c$: redistribute
- Avalanche dynamics

Power-law distributions:

$$P(s) \propto s^{-\tau} \tag{14.46}$$

Variational principle: System minimizes dissipation while maximizing transport

Applications:

- Earthquakes
- Neural avalanches
- Evolution

14.4.4 Swarm Intelligence

Collective behavior from simple rules:

Agent action: $S_i = \int dt[\text{alignment} + \text{cohesion} + \text{separation}]$

Specifically:

$$S_i = \int dt \left[-J \sum_j \hat{\mathbf{n}}_i \cdot \hat{\mathbf{n}}_j + K \sum_j V(|\mathbf{r}_i - \mathbf{r}_j|) \right] \tag{14.47}$$

Emergent behaviors:

- Flocking/schooling
- Ant trail formation
- Collective decision making

Information-theoretic action:

$$S_{info} = \int dt[I(\text{individual}; \text{collective}) - C(\text{constraints})] \tag{14.48}$$

14.4.5 Network Self-organization

Complex networks evolve to optimize function:
Network action: $S_{network} = \int dt \left[\sum_{ij} A_{ij} f_{ij} - \sum_i g_i\right]$
where:

- A_{ij} = adjacency matrix
- f_{ij} = flow/communication
- g_i = node capacity

Optimization yields:

- Scale-free topology
- Small-world properties
- Modular structure

Examples:

- Neural networks
- Vascular systems
- Social networks

14.4.6 Variational Principles in Evolution

Evolution as optimization process:
Fitness landscape: W(genotype)
Population action: $S_{evol} = \int dt \sum_i p_i[\ln W_i - \mu_{ij}(p_j - p_i)]$
where μ_{ij} are mutation rates.
Fisher's fundamental theorem:

$$\frac{d\bar{W}}{dt} = \text{Var}(W) \tag{14.49}$$

Price equation from variational principle:

$$\Delta \bar{z} = \frac{\text{Cov}(w, z)}{\bar{w}} + \frac{E(w \cdot \Delta z)}{\bar{w}} \tag{14.50}$$

Applications:

- Adaptive dynamics
- Evolutionary game theory
- Major transitions

14.5 Summary

The principle of least action extends remarkably into the realm of complex systems, providing unifying insights across scales and disciplines:

1. **Thermodynamic Foundations:** Classical thermodynamics emerges from action principles:
    - Maximum entropy for equilibrium states
    - Fluctuation theorems from path actions
    - Information-thermodynamic connections
2. **Non-equilibrium Extensions:** Irreversible processes follow variational principles:
    - Onsager's reciprocity from time-reversal symmetry
    - Minimum/maximum entropy production principles
    - Stochastic thermodynamics for small systems
3. **Dissipative Structures:** Order emerges from non-equilibrium:
    - Pattern formation in reaction-diffusion systems
    - Convective instabilities and turbulence
    - Biological organization through energy flux
4. **Self-organization:** Complex systems spontaneously order:
    - Emergent properties from collective dynamics
    - Critical phenomena without fine-tuning
    - Evolution as variational optimization

The action principle in complex systems reveals several profound insights:

Universality: Despite their diversity, complex systems share common organizational principles expressible through action functionals.

Emergence: Macroscopic order parameters and their dynamics emerge from microscopic actions through coarse-graining and constraint.

Optimization: Systems self-organize to extremize relevant functionals—entropy, information, fitness—subject to constraints.

Prediction: Variational principles provide predictive power even in complex, multi-scale systems where detailed microscopic description is impossible.

Design: Understanding these principles enables design of self-organizing systems, from materials to algorithms.

As we push into new frontiers—synthetic biology, artificial intelligence, quantum many-body systems—the principle of least action continues to guide our understanding. It reveals that beneath the apparent complexity lies deep simplicity: systems evolve along paths that extremize appropriate action functionals. This insight, which began with Fermat's light rays and Hamilton's mechanics, now illuminates the emergence of life, mind, and perhaps even consciousness itself from the fundamental laws of physics.

Chapter 15

Applications in Cosmology

Cosmology—the study of the universe as a whole—represents perhaps the grandest application of the principle of least action. From the Big Bang to the far future, from quantum fluctuations that seeded galaxies to the mysterious dark energy driving cosmic acceleration, the evolution of the universe follows from extremizing appropriate action functionals. The cosmological principle, asserting the universe's homogeneity and isotropy on large scales, combined with general relativity's action principle, yields a remarkably successful description of cosmic history.

This chapter explores how the principle of least action illuminates cosmology. We'll see how the Friedmann equations governing cosmic expansion emerge from the Einstein-Hilbert action, how inflation solves cosmological puzzles through a scalar field action, and how dark energy might arise from modifications to Einstein's theory. Throughout, the action principle not only describes what we observe but constrains possible cosmological scenarios, guiding our search for a complete theory of the universe.

15.1 Cosmological Action

15.1.1 The Friedmann-Lemaître-Robertson-Walker Universe

The cosmological principle—that the universe is homogeneous and isotropic on large scales—dramatically constrains the metric. The most general form consistent with these symmetries is the FLRW metric:

$$ds^2 = -c^2dt^2 + a^2(t)\left[\frac{dr^2}{1-kr^2} + r^2(d\theta^2 + \sin^2\theta d\phi^2)\right] \tag{15.1}$$

where:

- $a(t)$ is the scale factor
- $k = 0, \pm 1$ represents spatial curvature (flat, closed, open)

This reduces the infinite degrees of freedom in the metric to a single function $a(t)$.

15.1.2 The Cosmological Action

The total action for cosmology combines gravity and matter:

$$S = S_{EH} + S_{matter} + S_\Lambda \tag{15.2}$$

The Einstein-Hilbert action with cosmological constant:

$$S_{EH} = \frac{c^3}{16\pi G}\int d^4x\sqrt{-g}(R - 2\Lambda) \tag{15.3}$$

For the FLRW metric:

$$\sqrt{-g} = a^3 \sin\theta \tag{15.4}$$

$$R = 6\left[\frac{\ddot{a}}{a} + \left(\frac{\dot{a}}{a}\right)^2 + \frac{k}{a^2}\right] \tag{15.5}$$

The gravitational action becomes:

$$S_{grav} = \frac{\pi c^3}{2G}\int dt\, a^3\left[-3\left(\frac{\dot{a}^2}{a^2} + \frac{k}{a^2}\right) + \Lambda\right] \tag{15.6}$$

15.1.3 Matter Action and Energy-Momentum

For a perfect fluid with energy density ρ and pressure p:

$$T^{\mu\nu} = (\rho + p/c^2)u^\mu u^\nu + pg^{\mu\nu} \tag{15.7}$$

where u^μ is the four-velocity.

The matter action for a perfect fluid:

$$S_{matter} = -\int d^4x\sqrt{-g}\,\rho(n,s) \tag{15.8}$$

where n is number density and s is entropy per particle.

For the FLRW metric with comoving coordinates:

$$S_{matter} = -\int dt\, a^3 \rho \tag{15.9}$$

15.1.4 Friedmann Equations from Variation

Varying the total action with respect to $a(t)$:

$$\frac{\delta S}{\delta a} = 0 \tag{15.10}$$

yields the Friedmann equations:

$$\left(\frac{\dot{a}}{a}\right)^2 = \frac{8\pi G}{3}\rho - \frac{kc^2}{a^2} + \frac{\Lambda c^2}{3} \tag{15.11}$$

$$\frac{\ddot{a}}{a} = -\frac{4\pi G}{3}\left(\rho + \frac{3p}{c^2}\right) + \frac{\Lambda c^2}{3} \tag{15.12}$$

These describe cosmic expansion entirely in terms of energy content.

15.1.5 Conservation Equations

The Bianchi identity $\nabla_\mu T^{\mu\nu} = 0$ gives:

$$\dot{\rho} + 3\frac{\dot{a}}{a}\left(\rho + \frac{p}{c^2}\right) = 0 \tag{15.13}$$

For various components:

- Matter ($p = 0$): $\rho \propto a^{-3}$
- Radiation ($p = \rho c^2/3$): $\rho \propto a^{-4}$
- Cosmological constant ($p = -\rho c^2$): $\rho = \text{constant}$

15.1.6 Cosmological Solutions

The Friedmann equations admit various solutions:

- Matter-dominated ($k = 0$, $\Lambda = 0$): $a(t) = (t/t_0)^{2/3}$
- Radiation-dominated: $a(t) = (t/t_0)^{1/2}$
- De Sitter (Λ-dominated): $a(t) = \exp(Ht)$ where $H = \sqrt{\Lambda c^2/3}$

The current universe transitions from radiation to matter to dark energy domination.

15.2 Inflation and Principle of Least Action

15.2.1 The Inflationary Paradigm

Inflation—a period of exponential expansion in the early universe—solves several cosmological puzzles:

- Horizon problem: Why is the CMB uniform?
- Flatness problem: Why is $\Omega \approx 1$?
- Monopole problem: Where are the relics?

The key is a scalar field ϕ (inflaton) with appropriate action:

$$S = \int d^4x\sqrt{-g}\left[\frac{1}{2}\partial_\mu\phi\partial^\mu\phi - V(\phi)\right] \tag{15.14}$$

15.2.2 Slow-Roll Inflation

For successful inflation, the potential $V(\phi)$ must be flat enough. The slow-roll parameters:

$$\epsilon = \frac{M_p^2}{2}\left(\frac{V'}{V}\right)^2 \tag{15.15}$$

$$\eta = M_p^2\frac{V''}{V} \tag{15.16}$$

where $M_p = \sqrt{c^3\hbar/G}$ is the Planck mass.

Inflation requires $\epsilon, |\eta| \ll 1$, giving:

- Exponential expansion: $a(t) \propto \exp(Ht)$
- Nearly constant $H \approx \sqrt{V/3M_p^2}$
- Duration: $N = \int H dt \gtrsim 60$ e-folds

15.2.3 Quantum Fluctuations During Inflation

Quantum fluctuations of the inflaton field:

$$\langle \delta\phi_k \delta\phi_{k'} \rangle = (2\pi)^3 \delta^3(k + k') P_\phi(k) \tag{15.17}$$

where the power spectrum:

$$P_\phi(k) = \left(\frac{H}{2\pi}\right)^2 \bigg|_{k=aH} \tag{15.18}$$

These fluctuations:

- Are stretched to cosmic scales
- Seed structure formation
- Generate primordial gravitational waves

The action for perturbations:

$$S_{pert} = \frac{1}{2} \int d^4x \, a^3 \left[(\delta\dot{\phi})^2 - \frac{(\nabla\delta\phi)^2}{a^2} - V''(\delta\phi)^2 \right] \tag{15.19}$$

15.2.4 Inflationary Observables

The primordial power spectra:

$$\text{Scalar perturbations: } P_s(k) = \frac{1}{2M_p^2 \epsilon} \left(\frac{H}{2\pi}\right)^2 \bigg|_{k=aH} \tag{15.20}$$

$$\text{Tensor perturbations: } P_t(k) = 16 \left(\frac{H}{2\pi M_p}\right)^2 \bigg|_{k=aH} \tag{15.21}$$

Observable parameters:

- Spectral index: $n_s = 1 - 6\epsilon + 2\eta$

- Tensor-to-scalar ratio: $r = 16\epsilon$
- Running: $dn_s/d\ln k$

Current constraints:

- $n_s = 0.965 \pm 0.004$
- $r < 0.06$ (95% CL)

15.2.5 Inflationary Models

Different potentials yield distinct predictions:

Chaotic inflation: $V = \frac{1}{2}m^2\phi^2$

- $n_s = 1 - 2/N$
- $r = 8/N$

Natural inflation: $V = \Lambda^4[1 + \cos(\phi/f)]$

- Axionic origin
- Protected by shift symmetry

Starobinsky inflation: From R^2 gravity

$$S = \int d^4x\sqrt{-g}\left[\frac{M_p^2 R}{2} + \frac{R^2}{6M^2}\right] \tag{15.22}$$

- $n_s = 1 - 2/N$
- $r = 12/N^2$

15.2.6 Multifield Inflation

With multiple scalar fields ϕ^i:

$$S = \int d^4x\sqrt{-g}\left[\frac{1}{2}G_{ij}\partial_\mu\phi^i\partial^\mu\phi^j - V(\phi^i)\right] \tag{15.23}$$

New phenomena:

- Non-Gaussianity from field interactions
- Isocurvature perturbations
- Complex inflationary trajectories

The effective action for adiabatic and entropy modes reveals rich dynamics.

15.3 Dark Energy and Action Principle

15.3.1 The Cosmological Constant Problem

Observations indicate accelerated expansion, implying dark energy with:

$$\rho_{DE} \approx 0.7\rho_{crit} \approx 10^{-47}\ \text{GeV}^4 \tag{15.24}$$

The cosmological constant Λ is the simplest explanation, but:

- Quantum corrections: $\langle \rho_{vac} \rangle \sim M_p^4 \sim 10^{76}\ \text{GeV}^4$
- Fine-tuning: 120 orders of magnitude!

The action perspective:

$$S = \int d^4x\sqrt{-g}\left[\frac{R}{16\pi G} - \Lambda_{bare} - \Lambda_{quantum}\right] \tag{15.25}$$

Somehow $\Lambda_{bare} + \Lambda_{quantum} \approx 10^{-120} M_p^4$.

15.3.2 Dynamical Dark Energy

Scalar field dark energy (quintessence):

$$S = \int d^4x\sqrt{-g}\left[\frac{1}{2}\partial_\mu\phi\partial^\mu\phi - V(\phi)\right] \tag{15.26}$$

For cosmic acceleration:

- $V(\phi)$ dominates over kinetic energy
- $V(\phi)$ decreases slowly
- $w = p/\rho = (\dot{\phi}^2/2 - V)/(\dot{\phi}^2/2 + V) \approx -1$

Tracker solutions:

$$V(\phi) = M^{4+n}\phi^{-n} \tag{15.27}$$

attract solutions regardless of initial conditions.

15.3.3 Modified Gravity Approaches

Instead of dark energy, modify gravity:

$f(R)$ **gravity:**

$$S = \int d^4x\sqrt{-g}\left[\frac{f(R)}{16\pi G} + \mathcal{L}_{matter}\right] \tag{15.28}$$

Example: $f(R) = R + \alpha R^2$

- Yields late-time acceleration
- Must satisfy solar system constraints

DGP (brane-world):

$$S = \int_{brane} d^4x\sqrt{-g}\frac{R}{16\pi G} + \int_{bulk} d^5x\sqrt{-G}\frac{R_5}{16\pi G_5} \tag{15.29}$$

Leads to "self-accelerating" solution without dark energy.

15.3.4 Effective Field Theory of Dark Energy

Model-independent approach:

$$S = \int d^4x\sqrt{-g}\left[\frac{M_p^2}{2}f(t)R + \Lambda(t) - c(t)g^{00} + \frac{M_2^4(t)}{2}(\delta g^{00})^2 + ...\right] \tag{15.30}$$

where functions $f(t)$, $\Lambda(t)$, $c(t)$, $M_2(t)$ parametrize deviations from ΛCDM.

This encompasses:

- Quintessence
- k-essence
- Galileons
- Massive gravity

15.3.5 Observational Constraints

Dark energy probed through:

- Geometry: Distance-redshift relation

$$d_L(z) = (1+z)\int_0^z \frac{dz'}{H(z')} \tag{15.31}$$

- Growth: Structure formation

$$\ddot{\delta} + 2H\dot{\delta} - 4\pi G\rho\delta = 0 \tag{15.32}$$

The action determines both background evolution and perturbations.

Current constraints:

- $w = -1.03 \pm 0.03$ (consistent with Λ)
- No evidence for $w(z)$ evolution

15.3.6 Quantum Effects and Vacuum Energy

The effective action including quantum corrections:

$$\Gamma_{eff} = S_{classical} + \hbar\Gamma^{(1)} + \hbar^2\Gamma^{(2)} + ... \tag{15.33}$$

One-loop contribution:

$$\Gamma^{(1)} = \frac{1}{2}\text{Tr}\ \ln\bigl(\Box + m^2 + \xi R\bigr) \tag{15.34}$$

This generates:

- Running of G and Λ
- R^2 and higher-derivative terms
- Non-local corrections

The coincidence problem: Why $\rho_{DE} \sim \rho_{matter}$ today?

15.4 Modified Gravity and Alternative Actions

15.4.1 Beyond Einstein-Hilbert

The most general action for metric gravity:

$$S = \int d^4x\sqrt{-g}\, f(R, R_{\mu\nu}R^{\mu\nu}, R_{\mu\nu\rho\sigma}R^{\mu\nu\rho\sigma}, \Box R, \Box^2 R, ...) \tag{15.35}$$

Lovelock's theorem: In 4D, only R gives second-order field equations. This motivates:

- Higher dimensions
- Additional fields
- Non-local modifications

15.4.2 Scalar-Tensor Theories

Brans-Dicke theory:

$$S = \int d^4x\sqrt{-g}\left[\frac{\phi R}{16\pi} - \frac{\omega\nabla_\mu\phi\nabla^\mu\phi}{\phi} - V(\phi)\right] \tag{15.36}$$

Generalizes to Horndeski theory—most general scalar-tensor theory with second-order equations:

$$S = \sum_{i=2}^{5}\int d^4x\sqrt{-g}\,\mathcal{L}_i \tag{15.37}$$

where $\mathcal{L}_i$ are specific combinations of ϕ, $X = -\partial_\mu\phi\partial^\mu\phi/2$, and curvature.

15.4.3 Massive Gravity

Give the graviton a mass:

$$S = M_p^2\int d^4x\sqrt{-g}\left[\frac{R}{2} - m^2 U(g, f)\right] \tag{15.38}$$

where $f_{\mu\nu}$ is a reference metric and U generates mass terms.

Ghost-free construction (dRGT):

$$U = \alpha_1[K] + \alpha_2[K]^2 - \alpha_3[K^3] \tag{15.39}$$

where $K^\mu_\nu = \delta^\mu_\nu - \sqrt{g^{\mu\rho} f_{\rho\nu}}$.

Phenomenology:

- Vainshtein screening near sources
- Modified cosmology
- Gravitational wave propagation

15.4.4 Einstein-Aether and Lorentz Violation

Introduce a preferred frame via vector field:

$$S = \int d^4x\sqrt{-g}\left[-\frac{R}{16\pi G} + K^{\alpha\beta\mu\nu}\nabla_\alpha u_\mu \nabla_\beta u_\nu + \lambda(u_\mu u^\mu + 1)\right] \tag{15.40}$$

where u^μ is a unit timelike vector.

Consequences:

- Modified gravitational wave speed
- Preferred frame effects
- Black hole hair

15.4.5 Hořava-Lifshitz Gravity

Power-counting renormalizable by breaking Lorentz invariance:

$$S = \int dt d^3x \sqrt{h} N \left[K_{ij}K^{ij} - \lambda K^2 + \mathcal{L}_V\right] \tag{15.41}$$

where $\mathcal{L}_V$ contains spatial derivatives up to $2[D]$ order.

Features:

- UV complete candidate
- Anisotropic scaling: $t \to b^z t$, $x \to bx$
- Recovers GR in IR

15.4.6 Emergent Gravity

Gravity as emergent phenomenon:

Entropic gravity (Verlinde):

$$F = T\nabla S \tag{15.42}$$

where surface degrees of freedom generate gravitational force.

Induced gravity (Sakharov):

$$S_{eff} = -\frac{\hbar}{2}\text{Tr}\ \ln(\Box + m^2) \approx \int d^4x\sqrt{-g}\frac{R}{16\pi G_{induced}} \tag{15.43}$$

Gravity emerges from quantum fluctuations of matter.

AdS/CFT correspondence:

$$S_{bulk}[g] = S_{boundary}[CFT] \tag{15.44}$$

Gravity in bulk emerges from boundary field theory.

Chapter 16

Unifying Theories

The quest for a unified theory of nature—combining quantum mechanics, general relativity, and all fundamental forces—represents perhaps the most ambitious application of the principle of least action. At the Planck scale where quantum gravity effects become important, our familiar notions of space, time, and even the action itself require radical revision. Yet remarkably, the variational principle continues to guide theoretical development, suggesting forms for the fundamental action that might describe nature at its deepest level.

This chapter explores how modern attempts at unification employ and extend the action principle. String theory replaces point particles with extended objects, requiring new forms of actions. M-theory unifies different string theories through dualities, with branes playing crucial roles. Loop quantum gravity directly quantizes the Einstein-Hilbert action using new variables. Throughout these approaches, the principle of least action not only survives but becomes even more central, constraining the possible forms of unified theories and revealing unexpected connections between seemingly different formulations.

16.1 String Theory and Effective Actions

16.1.1 The String Action

String theory's foundational insight is replacing point particles with one-dimensional extended objects. The worldsheet swept out by a string is parametrized by (τ, σ), with embedding coordinates $X^\mu(\tau, \sigma)$ in spacetime.

The Nambu-Goto action equals the worldsheet area:

$$S_{NG} = -T \int d\tau d\sigma \sqrt{-\det h_{ab}} \tag{16.1}$$

where $h_{ab} = \partial_a X^\mu \partial_b X_\mu$ is the induced metric and $T = 1/(2\pi\alpha')$ is the string tension.

The equivalent Polyakov action:

$$S_P = -\frac{T}{2} \int d\tau d\sigma \sqrt{-h} h^{ab} \partial_a X^\mu \partial_b X_\mu \tag{16.2}$$

introduces an auxiliary worldsheet metric h_{ab}, making the action quadratic in X^μ.

16.1.2 String Spectrum and Constraints

Quantizing the string reveals a tower of states:

Open strings:

- Massless: gauge bosons (photon, gluons)
- Massive: $\alpha' m^2 = N - 1$

Closed strings:

- Massless: graviton, dilaton, B-field
- Massive: $\alpha' m^2 = 4(N + \tilde{N} - 2)$

Critical dimension: $D = 26$ (bosonic), $D = 10$ (superstring)
Constraints from Weyl invariance:

- Vanishing β-functions $\rightarrow$ Einstein equations
- Consistency requires specific dimensions
- Anomaly cancellation constrains gauge groups

16.1.3 Superstring Actions

Supersymmetry on the worldsheet requires fermionic coordinates ψ^μ:

$$S = -\frac{T}{2} \int d^2\sigma \left[\partial_a X^\mu \partial^a X_\mu + i\bar{\psi}^\mu \rho^a \partial_a \psi_\mu\right] \tag{16.3}$$

where ρ^a are worldsheet gamma matrices.

Five consistent superstring theories in 10D:

- Type I: Open and closed, SO(32) gauge group
- Type IIA/IIB: Closed only, different chiralities
- Heterotic SO(32), $E_8 \times E_8$: Hybrid construction

The effective spacetime action at low energies:

$$S_{eff} = \frac{1}{2\kappa^2} \int d^{10}x\sqrt{-G}e^{-2\Phi} \left[R + 4(\nabla\Phi)^2 - \frac{1}{12}H^2_{\mu\nu\rho} + ...\right] \tag{16.4}$$

where Φ is the dilaton and $H = dB$ is the field strength of the Kalb-Ramond field.

16.1.4 Compactification and Effective Actions

To connect with 4D physics, compactify 6 dimensions:

$$M^{10} = M^4 \times K_6 \tag{16.5}$$

The 4D effective action:

$$S_{4D} = \int d^4x\sqrt{-g} \left[\frac{M_p^2 R}{2} - \frac{1}{4}F^2_{IJ} - K_{i\bar{j}}\partial_\mu z^i \partial^\mu \bar{z}^{\bar{j}} - V(z, \bar{z})\right] \tag{16.6}$$

where:

- z^i are moduli fields from K_6 geometry
- $K_{i\bar{j}}$ is the Kähler metric on moduli space
- V is the potential from fluxes

Calabi-Yau compactifications preserve $\mathcal{N} = 1$ supersymmetry:

- $h^{1,1}$ vector multiplets
- $h^{2,1} + 1$ chiral multiplets
- Yukawa couplings from triple intersections

16.1.5 D-Branes and Open String Effective Actions

D-branes are surfaces where open strings end. The low-energy effective action on a Dp-brane:

$$S_{DBI} = -T_p \int d^{p+1}\xi \, e^{-\Phi} \sqrt{-\det(g_{ab} + 2\pi\alpha' F_{ab})} \tag{16.7}$$

This Dirac-Born-Infeld action:

- Reduces to Maxwell at weak fields
- Has maximum electric field $E_{max} = 1/(2\pi\alpha')$
- Describes non-linear electrodynamics

Including fermions gives the full supersymmetric action.

16.1.6 String Dualities and M-Theory

Different string theories are related by dualities:

T-duality: $R \to \alpha'/R$

- Exchanges momentum and winding modes
- IIA $\leftrightarrow$ IIB under T-duality

S-duality: $g_s \to 1/g_s$

- Strong-weak coupling duality
- Type I $\leftrightarrow$ SO(32) heterotic

U-duality: Combines S and T

These suggest all string theories are limits of a single theory—M-theory in 11D.

16.2 M-Theory and Branes

16.2.1 Eleven-Dimensional Supergravity

The low-energy limit of M-theory is 11D supergravity:

$$S_{11} = \frac{1}{2\kappa_{11}^2} \int d^{11}x\sqrt{-G}\left[R - \frac{1}{48}F_{MNPQ}^2 - \frac{1}{144^2}\varepsilon^{M_1 \ldots M_{11}} F_{M_1 \ldots M_4} F_{M_5 \ldots M_8} A_{M_9 M_{10} M_{11}}\right] \quad (16.8)$$

where $F = dA$ is the field strength of the 3-form potential.

This unique action:

- Has no dimensionless coupling
- Reduces to IIA string theory upon compactification
- Contains M2 and M5-branes as solitons

16.2.2 M2-Branes

The worldvolume action for M2-branes (supermembranes):

$$S_{M2} = -T_2 \int d^3\sigma \sqrt{-\det(h_{ij})} + \frac{T_2}{6} \int A_3 \quad (16.9)$$

where h_{ij} is the induced metric and A_3 is the pullback of the 3-form.

Multiple M2-branes are described by:

- BLG theory: Based on 3-algebras
- ABJM theory: $\mathcal{N} = 6$ Chern-Simons matter theory

The ABJM action:

$$S_{ABJM} = \int d^3x \left[\text{Tr}(D_\mu Z^I D^\mu \bar{Z}_I) + \text{fermions} + \frac{k}{4\pi}\varepsilon^{\mu\nu\rho}\text{Tr}(A_\mu \partial_\nu A_\rho + \ldots)\right] \quad (16.10)$$

16.2.3 M5-Branes

M5-branes support a self-dual 3-form field strength $H = dB + C_3$:

$$dH = 0, \quad *H = H \tag{16.11}$$

The non-linear action (schematically):

$$S_{M5} = -T_5 \int d^6\sigma \sqrt{1 + H^2} + T_5 \int C_6 \tag{16.12}$$

Multiple M5-branes lead to the mysterious (2,0) theory in 6D:

- No Lagrangian description known
- Contains self-dual fields
- Compactification yields 4D $\mathcal{N} = 2$ theories

16.2.4 Brane Intersections and Gauge Theories

Intersecting branes realize gauge theories:

D3-branes: $\mathcal{N} = 4$ SYM in 4D

$$S = \frac{1}{g_{YM}^2} \int d^4x \, \mathrm{Tr}[F_{\mu\nu}^2 + (D_\mu \Phi^i)^2 + [\Phi^i, \Phi^j]^2] \tag{16.13}$$

D4-D8 system: QCD-like theories

- D4s: color branes
- D8s: flavor branes
- Chiral symmetry breaking from D8 bending

Hanany-Witten setups: Brane configurations encoding field theory data

- Gauge group from brane numbers
- Matter from strings
- Superpotential from brane bending

16.2.5 AdS/CFT Correspondence

The holographic principle relates:

- Type IIB string theory on $\mathrm{AdS}_5 \times S^5$
- $\mathcal{N} = 4$ SYM in 4D

The correspondence:

$$Z_{string}[\phi_0] = \langle \exp\left(\int \phi_0 \mathcal{O}\right) \rangle_{CFT} \quad (16.14)$$

Gravity action in bulk ↔ Generating functional on boundary
Dictionary:

- Bulk fields ↔ Boundary operators
- Radial direction ↔ Energy scale
- Black holes ↔ Thermal states

Applications:

- Strong coupling calculations
- Quantum gravity in AdS
- Condensed matter systems

16.2.6 Higher Derivative Corrections

M-theory receives corrections to 11D supergravity:

$$S = S_{11} + l_p^3 \int d^{11}x\sqrt{-G}\left[c_1 R^4 + c_2(\nabla F)^2 F^2 + ...\right] + \mathcal{O}(l_p^6) \quad (16.15)$$

These corrections:

- Fixed by supersymmetry
- Related by dualities
- Encode quantum gravity effects

The complete M-theory action remains unknown, but constraints from:

- Supersymmetry
- Dualities
- Anomaly cancellation

severely restrict its form.

16.3 Loop Quantum Gravity

16.3.1 Canonical Formulation

Loop quantum gravity directly quantizes Einstein gravity using Hamiltonian methods. Starting with the Einstein-Hilbert action in first-order form:

$$S = \int dtd^3x \left[\dot{E}_i^a A_a^i - NH - N^a H_a \right] \tag{16.16}$$

where:

- E_i^a: Densitized triad (spatial metric)
- A_a^i: SU(2) connection (extrinsic curvature)
- N, N^a: Lapse and shift
- H, H_a: Hamiltonian and diffeomorphism constraints

16.3.2 Ashtekar Variables

The key insight: Use connection A_a^i as configuration variable instead of metric.

Canonical commutation relations:

$$\{A_a^i(x), E_j^b(y)\} = 8\pi\gamma G \delta_j^i \delta_a^b \delta^3(x-y) \tag{16.17}$$

where γ is the Barbero-Immirzi parameter.

Constraints become polynomial in A:

- Gauss: $D_a E_i^a = 0$
- Diffeomorphism: $F_{ab}^i E_i^b = 0$
- Hamiltonian: $\varepsilon_{ijk} E_i^a E_j^b F_{ab}^k / \sqrt{\det E} = 0$

16.3.3 Quantum Kinematics

Quantum states are functionals of connections $\Psi[A]$.

Key structures:

- Holonomies: $h_e[A] = \mathcal{P}\exp\left(\int_e A\right)$
- Wilson loops: $W_\gamma[A] = \mathrm{Tr}(h_\gamma[A])$
- Spin networks: $\Psi_\Gamma[A] = \prod_e \rho^{j_e}(h_e) \prod_v \iota_v$

The kinematical Hilbert space $\mathcal{H}_{kin}$ is spanned by spin network states:

- Edges labeled by SU(2) representations j
- Vertices by intertwiners ι

16.3.4 Quantum Geometry

Geometric operators have discrete spectra:

Area operator:

$$\hat{A}|j_e\rangle = 8\pi\gamma l_p^2\sqrt{j_e(j_e+1)}|j_e\rangle \tag{16.18}$$

Volume operator:

$$\hat{V}|\iota_v\rangle = \text{discrete spectrum depending on vertex valence} \tag{16.19}$$

This suggests:

- Space is discrete at Planck scale
- Minimum area $\sim l_p^2$
- Geometry emerges from quantum states

16.3.5 Dynamics and Constraints

The Hamiltonian constraint is problematic:

$$\hat{H}|\Psi\rangle = 0 \tag{16.20}$$

Challenges:

- Factor ordering ambiguities
- Anomalies in constraint algebra
- Defining dynamics

Approaches:

- Master constraint: $M = \int d^3x\sqrt{H^2 + H_a H^a}$
- Spin foam models: Path integral formulation
- Group field theory: Generating functionals

16.3.6 Spin Foam Models

The path integral approach sums over histories of spin networks:

$$Z = \sum_{\text{2-complex}} \prod_f A_f \prod_e A_e \prod_v A_v \tag{16.21}$$

where:

- Faces f: Labeled by representations
- Edges e: Carry intertwiners
- Vertices v: Encode local dynamics

The vertex amplitude for 4D gravity (EPRL/FK model):

$$A_v = \int \prod_e dg_e \prod_f \left[\Pi^{j_f}_{\gamma j_f}(g_f)\right] \tag{16.22}$$

Connection to classical gravity:

- Large j limit $\rightarrow$ Regge action
- Coherent states $\rightarrow$ Classical geometries

16.4 Non-perturbative Approaches

16.4.1 Asymptotic Safety

Weinberg's asymptotic safety scenario: Gravity might be non-perturbatively renormalizable with a UV fixed point.

The Wetterich equation for the effective action:

$$\partial_t \Gamma_k = \frac{1}{2} \mathrm{Tr} \left[(\Gamma_k^{(2)} + R_k)^{-1} \partial_t R_k \right] \tag{16.23}$$

where R_k is a momentum cutoff at scale k.

Evidence for fixed point:

- Einstein-Hilbert truncation: $g_* \approx 0.7$, $\lambda_* \approx 0.2$
- Higher derivative terms under control
- Finite number of relevant couplings

16.4.2 Causal Dynamical Triangulations

CDT provides a non-perturbative path integral for quantum gravity:

$$Z = \sum_{\text{triangulations}} \frac{1}{C_T} e^{-S_{Regge}[T]} \tag{16.24}$$

Key features:

- Causal structure preserved
- Foliation into spatial slices
- Euclidean rotation after sum

Results:

- 4D de Sitter universe emerges
- Spectral dimension runs: $D_s \to 2$ as $k \to \infty$
- Phase diagram with different geometries

16.4.3 Group Field Theory

GFT generates spin foams as Feynman diagrams:

$$S_{GFT} = \int \prod_i dg_i \left[K(g_1, ..., g_d)\phi(g_1, ..., g_d) + \frac{\lambda}{5!} \int \prod_j dg'_j V(g, g')\phi^5 \right] \tag{16.25}$$

where:

- ϕ: Fields on group manifold G^d
- K: Kinetic term (propagator)
- V: Vertex (interaction)

The partition function:

$$Z_{GFT} = \int D\phi\, e^{-S_{GFT}[\phi]} = \sum_\Gamma \lambda^{V_\Gamma} A_\Gamma \tag{16.26}$$

generates all spin foam amplitudes.

16.4.4 Causal Sets

The causal set approach: Spacetime is fundamentally discrete.

Postulates:

- Spacetime = locally finite poset
- Order relation = causal structure
- Number of elements = volume

The action:

$$S = \sum_{i<j<k<l} \alpha(i, j, k, l) \tag{16.27}$$

counts specific causal relations.

Dynamics via:

- Sequential growth (classical)
- Sum over causal sets (quantum)

Predictions:

- Lorentz violation at Planck scale
- Modified dispersion relations
- Discrete cosmology

16.4.5 Emergent Gravity Scenarios

Various proposals for gravity as emergent:

Entropic gravity:

- Gravity from entanglement entropy
- Einstein equations as thermodynamic identity
- AdS/CFT as prime example

Condensed matter analogues:

- Acoustic metric in fluids: $g^{\mu\nu}_{eff} \propto$ flow field
- Graphene: Emergent Dirac fermions
- Lessons for quantum gravity

Tensor networks:

- MERA as holographic RG
- Entanglement = geometry
- Quantum error correction in AdS/CFT

16.4.6 Swampland Constraints

Not all effective field theories couple consistently to quantum gravity.

Swampland criteria:

- Weak gravity conjecture: $q \geq m/M_p$
- Distance conjecture: Infinite distance $\rightarrow$ massless tower
- de Sitter conjecture: $|\nabla V| \geq cV/M_p$

These constrain:

- Possible compactifications
- Inflationary models
- Dark energy solutions

The landscape/swampland program uses consistency to constrain physics.

Part IV

Numerical Methods and Practical Applications

Part V explores how the principle of least action translates into practical computational methods and real-world applications. While the mathematical beauty of variational principles is compelling, their true power emerges when applied to solve concrete problems—from designing materials with specific properties to optimizing complex engineering systems. Modern computational resources allow us to tackle problems far beyond analytical solution, yet the underlying variational structure continues to guide algorithm design and ensure physical fidelity.

We begin with numerical techniques that directly implement variational principles, showing how discretization schemes can preserve the geometric structure of mechanics. We then explore how the action principle inspires optimization algorithms and enables efficient simulation of complex systems. Finally, we examine engineering applications where optimal control theory—deeply rooted in variational principles—guides the design of everything from spacecraft trajectories to robotic systems. Throughout, we'll see that numerical implementation of the action principle requires careful attention to preserve the physical and mathematical structures that make these principles so powerful.

Chapter 17

Numerical Techniques

The transition from analytical to numerical solutions represents a crucial step in applying the principle of least action to real-world problems. While closed-form solutions exist only for highly idealized systems, numerical methods allow us to tackle the full complexity of nature—from protein folding to galaxy formation, from quantum chemistry to turbulent flows. The key insight is that successful numerical methods must respect the underlying variational structure, preserving conservation laws and geometric properties that the action principle guarantees.

This chapter explores how variational principles guide the development of numerical algorithms. We'll see how finite element methods naturally arise from discretizing action functionals, how Monte Carlo techniques evaluate path integrals in high dimensions, and how modern optimization algorithms draw inspiration from physical minimization principles. Throughout, the emphasis is on methods that not only solve equations but preserve the deep structure that makes the action principle so fundamental to physics.

17.1 Variational Methods in Numerical Computation

17.1.1 Discretization of Action Functionals

The continuous action $S[q(t)] = \int L(q, \dot{q}, t)dt$ must be discretized for numerical computation. The simplest approach divides time into N

intervals of size $h = (t_f - t_0)/N$:

$$S_{discrete} = h \sum_{i=0}^{N-1} L\left(q_i, \frac{q_{i+1} - q_i}{h}, t_i\right) \tag{17.1}$$

This immediately raises questions:

- How to approximate $\dot{q}$?
- Where to evaluate L?
- How to preserve conservation laws?

Different choices lead to different integration schemes with varying properties.

17.1.2 Symplectic Integrators

The key insight: preserve the symplectic structure of Hamiltonian mechanics.

For a Hamiltonian $H(q, p)$, exact evolution preserves the symplectic form:

$$\omega = \sum_i dp_i \wedge dq_i \tag{17.2}$$

A numerical method is symplectic if:

$$d\tilde{q}_i \wedge d\tilde{p}_i = dq_i \wedge dp_i \tag{17.3}$$

where $(\tilde{q}, \tilde{p})$ are the numerically evolved values.

The simplest symplectic integrator is the Störmer-Verlet method:

$$q_{n+1} = q_n + hp_n + \frac{h^2}{2} f(q_n) \tag{17.4}$$

$$p_{n+1} = p_n + \frac{h}{2}[f(q_n) + f(q_{n+1})] \tag{17.5}$$

This corresponds to a discrete action:

$$S_{SV} = h \sum_n \left[p_n \frac{q_{n+1} - q_n}{h} - H(q_n, p_n)\right] \tag{17.6}$$

17.1.3 Variational Integrators

Directly discretize the action and apply discrete variational principle:

$$\delta S_{discrete} = 0 \tag{17.7}$$

For the discrete Lagrangian $L_d(q_k, q_{k+1}) \approx \int_{t_k}^{t_{k+1}} L\,dt$:

$$\frac{\partial L_d}{\partial q_{k+1}}(q_{k-1}, q_k) + \frac{\partial L_d}{\partial q_k}(q_k, q_{k+1}) = 0 \tag{17.8}$$

This yields the discrete Euler-Lagrange equations.
Example—Harmonic oscillator:

$$L_d = h\left[\frac{(q_{k+1} - q_k)^2}{2h^2} - \frac{\omega^2 q_k^2}{2}\right] \tag{17.9}$$

Gives the discrete equation:

$$q_{k+1} - 2q_k + q_{k-1} = -h^2\omega^2 q_k \tag{17.10}$$

Key properties:

- Automatic conservation of discrete momentum maps
- Exact preservation of discrete symplectic form
- Bounded energy error for all time

17.1.4 Finite Element Methods

For field theories and PDEs, discretize space while keeping time continuous:

$$\phi(x, t) = \sum_i N_i(x)\phi_i(t) \tag{17.11}$$

where N_i are shape functions.
The action $S[\phi] = \int dt \int dx\, \mathcal{L}[\phi, \partial\phi/\partial t, \nabla\phi]$ becomes:

$$S[\phi_i] = \int dt\, L_{discrete}(\phi_i, \dot{\phi}_i) \tag{17.12}$$

where:

$$L_{discrete} = \sum_{ij} \frac{M_{ij}\dot{\phi}_i\dot{\phi}_j}{2} - \frac{K_{ij}\phi_i\phi_j}{2} \tag{17.13}$$

with mass matrix $M_{ij} = \int N_i N_j\, dx$ and stiffness matrix $K_{ij} = \int \nabla N_i \cdot \nabla N_j\, dx$.

17.1.5 Spectral Methods

For periodic systems or high accuracy requirements, expand in basis functions:

$$q(t) = \sum_{n=-N}^{N} c_n e^{in\omega t} \tag{17.14}$$

The action becomes algebraic in Fourier coefficients:

$$S[c_n] = \sum_n \left[\frac{(in\omega)^2 |c_n|^2}{2} - V(c_n) \right] \tag{17.15}$$

Advantages:

- Exponential convergence for smooth solutions
- Natural handling of periodic boundary conditions
- Efficient via FFT

Challenges:

- Aliasing for nonlinear terms
- Gibbs phenomena at discontinuities
- Limited to special geometries

17.1.6 Adaptive Methods

Vary discretization to maintain accuracy while minimizing computation:

h-adaptivity: Refine mesh where needed

- Error indicators from variational principle
- Conservation properties guide refinement

p-adaptivity: Vary polynomial order

- High order in smooth regions
- Low order near singularities

r-adaptivity: Move mesh points

- Minimize discretization error functional
- Preserve mesh topology

Combined hp-adaptive methods achieve exponential convergence.

17.2 Monte Carlo Methods and Path Integrals

17.2.1 Path Integral Formulation

The quantum propagator:

$$K(q_f, t_f; q_i, t_i) = \int \mathcal{D}[q] \exp(iS[q]/\hbar) \tag{17.16}$$

In imaginary time $\tau = it$ (Euclidean path integral):

$$K_E = \int \mathcal{D}[q] \exp(-S_E[q]/\hbar) \tag{17.17}$$

where S_E is the Euclidean action.

Discretization with N time slices:

$$K_E \approx \left(\frac{m}{2\pi\hbar\varepsilon}\right)^{N/2} \int \prod_{i=1}^{N-1} dq_i \exp(-S_{discrete}/\hbar) \tag{17.18}$$

17.2.2 Metropolis Algorithm

Generate path configurations with probability $P[q] \propto \exp(-S[q]/\hbar)$:

1. Start with path $q(\tau)$
2. Propose new path $q'(\tau) = q(\tau) + \delta q(\tau)$
3. Accept with probability $\min(1, \exp(-\Delta S/\hbar))$
4. Repeat to generate ensemble

Observables computed as ensemble averages:

$$\langle O \rangle = \frac{1}{N_{configs}} \sum_i O[q_i] \tag{17.19}$$

17.2.3 Importance Sampling

Direct sampling inefficient for small $\hbar$. Use importance sampling:

$$\langle O \rangle = \frac{\int \mathcal{D}[q] O[q] \exp(-S[q]/\hbar)}{\int \mathcal{D}[q] \exp(-S[q]/\hbar)} = \frac{\int \mathcal{D}[q] O[q] P[q]}{\int \mathcal{D}[q] P[q]} \tag{17.20}$$

where $P[q]$ is a probability distribution.

Optimal choice minimizes variance:

$$P_{opt}[q] \propto |O[q]| \exp(-S[q]/\hbar) \tag{17.21}$$

In practice, use approximations:

- Harmonic approximation around classical path
- Multilevel sampling
- Replica exchange

17.2.4 Quantum Monte Carlo

For many-body quantum systems:

Path Integral Monte Carlo (PIMC):

- Discretize imaginary time
- Sample all particle paths
- Exact for bosons, sign problem for fermions

Variational Monte Carlo (VMC):

- Trial wavefunction $\psi_T(R; \alpha)$
- Minimize $\langle H \rangle = \int |\psi_T|^2 H \psi_T \, dR / \int |\psi_T|^2 \, dR$
- Sample $|\psi_T|^2$ distribution

Diffusion Monte Carlo (DMC):

- Imaginary time Schrödinger equation
- $\partial\psi/\partial\tau = -(H - E_T)\psi$
- Projects out ground state

17.2.5 Lattice Field Theory

Discretize spacetime for field theories:

Lattice QCD:

$$S = \beta \sum_{\square} \left[1 - \frac{1}{3} \mathrm{Re}\,\mathrm{Tr}\, U_{\square} \right] + \sum_{x} \bar{\psi}(x) \not{D} \psi(x) \tag{17.22}$$

where $U_{\square}$ is the plaquette variable.

Monte Carlo updates:

- Gauge fields: Heatbath/overrelaxation
- Fermions: Hybrid Monte Carlo

Challenges:

- Critical slowing down near phase transitions
- Fermion determinant expensive
- Sign problem at finite density

17.2.6 Rare Event Sampling

For activated processes, standard Monte Carlo fails. Enhanced sampling methods:

Umbrella sampling: Add bias potential

$$S_{biased} = S + W(\xi) \tag{17.23}$$

where ξ is reaction coordinate.

Metadynamics: Build bias adaptively

$$W(\xi, t) = \sum_{t'<t} h \exp\left(-\frac{(\xi - \xi(t'))^2}{2\sigma^2} \right) \tag{17.24}$$

Transition path sampling: Sample reactive trajectories

$$P[\text{path}] \propto \exp(-S[\text{path}]/\hbar) \times h_A(q_0) \times h_B(q_N) \tag{17.25}$$

Forward flux sampling:

- Compute rate constants
- Define interfaces between states
- Sample trajectories between interfaces
- Combine to get total rate

17.3 Optimization Algorithms Inspired by Principle of Least Action

17.3.1 Gradient Descent as Overdamped Dynamics

Consider minimizing a function $f(x)$. The overdamped equation:

$$\dot{x} = -\nabla f(x) \tag{17.26}$$

follows from the action:

$$S = \int dt \left[\frac{\dot{x}^2}{2} + \dot{x} \cdot \nabla f(x)\right] \tag{17.27}$$

Discrete version gives gradient descent:

$$x_{n+1} = x_n - \eta \nabla f(x_n) \tag{17.28}$$

Improvements inspired by physics:

- Momentum (heavy ball method)
- Adaptive friction (AdaGrad, RMSprop)
- Preconditioning (natural gradient)

17.3.2 Hamiltonian Monte Carlo

Use Hamiltonian dynamics to explore probability distributions:

$$H(q, p) = \frac{p^2}{2m} + U(q) \tag{17.29}$$

where $U(q) = -\log P(q)$.

Algorithm:

1. Sample momentum $p \sim \mathcal{N}(0, m)$
2. Evolve (q, p) using symplectic integrator
3. Accept/reject based on energy change

Advantages:

- Explores high-dimensional spaces efficiently
- Reduces random walk behavior
- Preserves detailed balance

17.3.3 Simulated Annealing

Inspired by thermodynamics and metallurgy:

$$P(\text{state}) \propto \exp(-E(\text{state})/T) \tag{17.30}$$

Algorithm:

1. Start at high temperature T
2. Make random moves, accept via Metropolis
3. Slowly decrease $T \to 0$
4. System settles into low energy state

The cooling schedule is crucial:

- $T(t) = T_0/\log(1+t)$ guarantees convergence

Quantum annealing uses tunneling:

$$H(s) = (1-s)H_0 + sH_{problem} \tag{17.31}$$

Evolve $s : 0 \to 1$ adiabatically.

17.3.4 Genetic Algorithms and Evolution Strategies

View optimization as evolution with fitness = −action:

- Population of solutions $\{x_i\}$
- Fitness $f_i = \exp(-S[x_i]/T)$

Operations:

- Selection: Choose parents $\propto$ fitness
- Crossover: Combine parent solutions
- Mutation: Random perturbations

Evolution strategies use:

$$x_{n+1} = x_n + \sigma_n \mathcal{N}(0, I) \tag{17.32}$$

$$\sigma_{n+1} = \sigma_n \exp(\tau \mathcal{N}(0, 1)) \tag{17.33}$$

Self-adaptive step sizes from maximum entropy principle.

17.3.5 Variational Inference

Approximate intractable distribution $p(x)$ with tractable $q(x;\theta)$:

$$KL[q||p] = \int q(x) \log \frac{q(x)}{p(x)} dx \tag{17.34}$$

Minimize free energy:

$$F[q] = \langle E \rangle_q - S[q] \tag{17.35}$$

where $S[q] = -\int q \log q$ is entropy.

This connects to:

- Mean field theory
- Belief propagation
- Variational autoencoders

17.3.6 Physics-Informed Neural Networks

Incorporate physical laws into neural network training:

$$\text{Loss} = \text{MSE}_{data} + \lambda \int \left| \frac{\partial u}{\partial t} + N[u] \right|^2 dx\, dt \tag{17.36}$$

where $N[u]$ is the physical operator.

For variational problems:

$$\text{Loss} = \frac{\delta S[u_\theta]}{\delta u} \tag{17.37}$$

where u_θ is the neural network approximation.

Advantages:

- Automatic differentiation for derivatives
- Meshfree method
- Incorporates physical constraints

17.4 Applications in Molecular Simulation and Material Design

17.4.1 Molecular Dynamics

Newton's equations from the action:

$$S = \int dt \left[\sum_i \frac{m_i \dot{r}_i^2}{2} - V(r_1, ..., r_N) \right] \tag{17.38}$$

Standard MD:

$$r_i(t + \delta t) = r_i(t) + v_i(t)\delta t + \frac{f_i(t)\delta t^2}{2m_i} \tag{17.39}$$

$$v_i(t + \delta t) = v_i(t) + \frac{[f_i(t) + f_i(t + \delta t)]\delta t}{2m_i} \tag{17.40}$$

Enhancements:

- Multiple time stepping for different forces
- Constraint dynamics (SHAKE/RATTLE)
- Thermostats and barostats

17.4.2 Ab Initio Molecular Dynamics

Combine quantum electrons with classical nuclei:
Born-Oppenheimer MD:

1. Solve Schrödinger equation for electrons
2. Use forces on nuclei: $F_I = -\partial E_0 / \partial R_I$
3. Propagate nuclei classically

Car-Parrinello MD: Extended Lagrangian:

$$L = \sum_I \frac{M_I \dot{R}_I^2}{2} + \sum_i \mu \langle \dot{\psi}_i | \dot{\psi}_i \rangle - E[\{\psi_i\}, \{R_I\}] \tag{17.41}$$

Fictitious dynamics for orbitals keeps them near ground state.

17.4.3 Free Energy Calculations

Compute free energy differences:

$$\Delta F = -kT \log \frac{Z_B}{Z_A} \tag{17.42}$$

Methods: **Thermodynamic integration:**

$$\frac{\partial F}{\partial \lambda} = \left\langle \frac{\partial H}{\partial \lambda} \right\rangle_\lambda \tag{17.43}$$

$$\Delta F = \int_0^1 \left\langle \frac{\partial H}{\partial \lambda} \right\rangle_\lambda d\lambda \tag{17.44}$$

Free energy perturbation:

$$\Delta F = -kT \log\langle \exp(-\Delta H/kT) \rangle_A \tag{17.45}$$

Metadynamics: Build free energy surface

$$F(s) = -\int_0^t V(s, t')\, dt' + \text{const} \tag{17.46}$$

17.4.4 Crystal Structure Prediction

Find minimum of potential energy surface:

- Genetic algorithms: Evolve populations of structures
- Basin hopping: Random structure + local minimization
- Random structure search: Systematic exploration

Ab initio random structure search (AIRSS):

1. Generate random structures with constraints
2. Relax using DFT
3. Identify unique minima

Success stories:

- High-pressure phases of elements
- Novel materials (e.g., 2D materials)
- Drug polymorphs

17.4.5 Materials Design by Inverse Optimization

Design materials with target properties:

- Forward problem: Structure → Properties
- Inverse problem: Properties → Structure

Approaches:

- Gradient-based: ∂Property/∂Structure
- Genetic algorithms: Evolve toward target
- Machine learning: Learn structure-property maps

Example—Photonic crystals: Maximize bandgap via topology optimization

$$S = \int \left[\varepsilon(r)|E|^2 - |H|^2\right] dr \tag{17.47}$$

Optimize $\varepsilon(r)$ distribution.

17.4.6 Coarse-Graining and Multiscale Methods

Connect atomistic to continuum:

Force matching: Minimize:

$$\chi^2 = \sum_{config} \sum_i |F_i^{CG} - F_i^{mapped}|^2 \tag{17.48}$$

Relative entropy:

$$S_{rel} = \int P_{AA} \log \frac{P_{AA}}{P_{CG}} \, d\Gamma \tag{17.49}$$

Variational inference: Maximize likelihood of CG model given AA data

Applications:

- Polymer simulations
- Biomolecular systems
- Mesoscale phenomena

Chapter 18

Engineering and Optimal Control

The principle of least action finds its most direct engineering application in optimal control theory, where we seek control strategies that minimize cost functionals while respecting system dynamics. From guiding spacecraft to distant planets to controlling robotic manipulators with minimal energy, from designing efficient transportation systems to optimizing industrial processes, the variational approach provides both theoretical framework and practical algorithms. This final chapter demonstrates how the abstract principle that began with Fermat's light rays now guides the design of our most sophisticated technological systems.

Optimal control theory emerged in the 1950s with Pontryagin's maximum principle, which can be viewed as a generalization of the Euler-Lagrange equations to controlled systems. Modern applications span an enormous range: autonomous vehicles navigate using model predictive control based on action minimization, robotic systems plan motions that optimize task performance while minimizing energy, and even economic systems are analyzed through the lens of dynamic optimization. Throughout these applications, the principle of least action provides the unifying mathematical framework.

18.1 Pontryagin's Maximum Principle

18.1.1 The Optimal Control Problem

Consider a dynamical system with state $x(t) \in \mathbb{R}^n$ and control $u(t) \in U \subseteq \mathbb{R}^m$:

$$\dot{x} = f(x, u, t) \tag{18.1}$$

$$x(t_0) = x_0 \tag{18.2}$$

The goal is to find a control trajectory $u^*(t)$ that minimizes a cost functional:

$$J[u] = \phi(x(t_f), t_f) + \int_{t_0}^{t_f} L(x(t), u(t), t)\, dt \tag{18.3}$$

where:

- ϕ is the terminal cost
- L is the running cost (Lagrangian)
- t_f may be fixed or free

This generalizes the classical variational problem by allowing direct control inputs.

18.1.2 The Maximum Principle

Pontryagin's insight was to introduce the Hamiltonian:

$$H(x, u, \lambda, t) = \lambda^T f(x, u, t) - L(x, u, t) \tag{18.4}$$

where $\lambda(t) \in \mathbb{R}^n$ are costate variables (analogous to Lagrange multipliers).

The maximum principle states that optimal trajectories satisfy:

1. State equation: $\dot{x}^* = \partial H/\partial\lambda = f(x^*, u^*, t)$
2. Costate equation: $\dot{\lambda}^* = -\partial H/\partial x$
3. Optimality condition: $H(x^*, u^*, \lambda^*, t) = \max_{u \in U} H(x^*, u, \lambda^*, t)$
4. Transversality conditions at boundaries

The key difference from classical mechanics: we maximize (not just extremize) H with respect to u.

18.1.3 Connection to Hamilton-Jacobi Theory

Define the value function:

$$V(x,t) = \min_{u[t,t_f]} \left\{ \phi(x(t_f), t_f) + \int_t^{t_f} L(x(s), u(s), s)\, ds \right\} \tag{18.5}$$

This satisfies the Hamilton-Jacobi-Bellman equation:

$$-\frac{\partial V}{\partial t} = \min_u \left\{ L(x,u,t) + \left(\frac{\partial V}{\partial x}\right)^T f(x,u,t) \right\} \tag{18.6}$$

The costate variables are related to the value function:

$$\lambda = \frac{\partial V}{\partial x} \tag{18.7}$$

This connects optimal control to classical mechanics: the value function is the action!

18.1.4 Bang-Bang Control

When the control appears linearly in the Hamiltonian:

$$H = \lambda^T (A(x)x + B(x)u) - L_0(x) \tag{18.8}$$

The optimality condition gives:

$$u^* = \begin{cases} u_{max} & \text{if } \lambda^T B > 0 \\ u_{min} & \text{if } \lambda^T B < 0 \end{cases} \tag{18.9}$$

The control "bangs" between extremes. The switching function $\sigma = \lambda^T B$ determines transitions.

Example—Minimum time problem:

- $L = 1$ (minimize time)
- Double integrator: $\ddot{x} = u$, $|u| \leq 1$
- Optimal control switches at most once

18.1.5 Singular Control

When the switching function vanishes on an interval:

$$\sigma(t) \equiv 0 \quad \text{for } t \in [t_1, t_2] \tag{18.10}$$

The control is not determined by the maximum principle alone. Additional conditions:

- $\dot{\sigma} = 0$, $\ddot{\sigma} = 0$, ...
- Eventually, u appears explicitly, determining singular control

Applications:

- Rocket thrust programming
- Economic growth models
- Biological resource management

18.1.6 Numerical Methods

Solving optimal control problems numerically:

Indirect methods: Solve the two-point boundary value problem

- State equations forward
- Costate equations backward
- Match at boundaries

Direct methods: Discretize and optimize

- Collocation: Satisfy dynamics at points
- Shooting: Parameterize control
- Pseudospectral: Expand in basis functions

Modern software packages:

- GPOPS-II (pseudospectral)
- ACADO (multiple methods)
- CasADi (symbolic + numeric)

18.2 Optimal Control and Principle of Least Action

18.2.1 Variational Formulation of Optimal Control

The optimal control problem is a generalized variational problem. Define the augmented action:

$$S[x, u, \lambda] = \phi(x(t_f), t_f) + \int_{t_0}^{t_f} [L(x, u, t) + \lambda^T(\dot{x} - f(x, u, t))]\, dt \quad (18.11)$$

Varying with respect to x, u, and λ:

$$\delta S/\delta x: \quad \dot{\lambda} = -\frac{\partial L}{\partial x} + \left(\frac{\partial f}{\partial x}\right)^T \lambda \quad (18.12)$$

$$\delta S/\delta u: \quad 0 = \frac{\partial L}{\partial u} - \left(\frac{\partial f}{\partial u}\right)^T \lambda \quad (18.13)$$

$$\delta S/\delta \lambda: \quad \dot{x} = f(x, u, t) \quad (18.14)$$

These are exactly Pontryagin's conditions!

18.2.2 Control as Generalized Force

View the controlled system as:

$$\frac{d}{dt}\left(\frac{\partial T}{\partial \dot{x}}\right) - \frac{\partial T}{\partial x} = Q + Bu \quad (18.15)$$

where:

- T is kinetic energy
- Q are conservative forces
- Bu are control forces

The action becomes:

$$S = \int \left[T - V + u^T B^T \left(\frac{\partial T}{\partial \dot{x}}\right)\right] dt - \int \ell(u)\, d\dot{t} \quad (18.16)$$

This reveals control as choosing non-conservative forces to minimize total action.

18.2.3 Model Predictive Control

MPC solves a finite-horizon optimal control problem repeatedly:
At time t, solve:

$$\min_{u[t,t+T]} \int_t^{t+T} L(x,u)\,ds + V(x(t+T)) \tag{18.17}$$

Apply only first portion, then re-solve.
This is a receding-horizon implementation of the action principle:

- Local in time
- Feedback through re-optimization
- Handles constraints naturally

Connection to physics:

- Finite propagation speed
- Causality constraints
- Local action principles

18.2.4 Stochastic Optimal Control

For systems with noise:

$$dx = f(x,u)dt + \sigma(x)dW \tag{18.18}$$

The cost becomes an expectation:

$$J = \mathbb{E}\left[\phi(x(t_f)) + \int L(x,u)dt\right] \tag{18.19}$$

The Hamilton-Jacobi-Bellman equation becomes:

$$-\frac{\partial V}{\partial t} = \min_u \left\{ L + f^T \frac{\partial V}{\partial x} + \frac{1}{2}\mathrm{Tr}\left(\sigma\sigma^T \frac{\partial^2 V}{\partial x^2}\right) \right\} \tag{18.20}$$

This is the stochastic action principle, related to:

- Path integral formulation
- Large deviation theory
- Fluctuation theorems

18.2.5 Inverse Optimal Control

Given observed trajectories, infer the cost function being optimized:

Assume: $J = \int w^T \phi(x, u) dt$

where ϕ are features and w are unknown weights.

From optimality conditions:

$$\nabla_u H = w^T \nabla_u \phi - \lambda^T \nabla_u f = 0 \tag{18.21}$$

This linear equation in w can be solved from data.

Applications:

- Learning from demonstration
- Understanding biological motion
- Economic behavior modeling

18.2.6 Connections to Reinforcement Learning

RL seeks policy $\pi(u|x)$ maximizing expected reward:

$$J = \mathbb{E}\left[\sum_t \gamma^t r(x_t, u_t)\right] \tag{18.22}$$

The Bellman equation:

$$V(x) = \max_u \{r(x, u) + \gamma \mathbb{E}[V(x')]\} \tag{18.23}$$

is a discrete-time Hamilton-Jacobi equation.

Policy gradient methods:

$$\nabla_\theta J = \mathbb{E}[\nabla_\theta \log \pi_\theta(u|x) Q(x, u)] \tag{18.24}$$

connect to the calculus of variations.

Deep RL approximates value functions and policies with neural networks, implementing approximate dynamic programming.

18.3 Applications in Robotics and Autonomous Systems

18.3.1 Robot Motion Planning

For a robot with configuration $q \in \mathbb{R}^n$, find a trajectory minimizing:

$$J = \int_0^T \left[\frac{1}{2}\dot{q}^T M(q)\dot{q} + \ell(q, \dot{q})\right] dt \tag{18.25}$$

subject to:

- Dynamics: $M(q)\ddot{q} + C(q, \dot{q})\dot{q} + g(q) = \tau$
- Constraints: $h(q) \geq 0$ (obstacles)
- Limits: $\tau_{min} \leq \tau \leq \tau_{max}$

Methods:

- Trajectory optimization: Discretize and solve NLP
- Sampling-based: RRT*, PRM* minimize path cost
- Optimal control: DDP, iLQG for dynamics

18.3.2 Legged Locomotion

Walking/running as hybrid optimal control:

- Continuous dynamics: $M(q)\ddot{q} + C(q, \dot{q})\dot{q} + g(q) = B\tau + J^T(q)F$
- Contact constraints: $\phi(q) \geq 0$, $F \geq 0$, $F^T\phi = 0$

Cost function typically includes:

- Energy: $\int \tau^T\tau \, dt$
- Stability: $\int (x - x_{ref})^T Q(x - x_{ref}) \, dt$
- Task: Terminal constraints

The action principle yields:

- Optimal ground reaction forces
- Efficient gait patterns
- Robust locomotion strategies

18.3.3 Manipulation and Grasping

Optimal grasping as variational problem:

$$\min \int [F^T K F + \lambda(\text{stability})] \, dt \tag{18.26}$$

subject to:

- Force balance: $Gf = w_{ext}$
- Friction constraints: $|f_t| \leq \mu f_n$
- Joint limits

G is the grasp matrix relating contact forces to object wrench. The action formulation naturally handles:

- Multi-contact scenarios
- Dynamic manipulation
- Human-robot interaction

18.3.4 Autonomous Vehicle Control

Self-driving cars solve receding-horizon optimal control:

$$\min \sum_k [(x_k - x_{ref})^T Q (x_k - x_{ref}) + u_k^T R u_k] \tag{18.27}$$

subject to:

- Vehicle dynamics (bicycle model)
- Road constraints (lanes)
- Obstacle avoidance
- Traffic rules

The variational structure enables:

- Smooth trajectory generation
- Constraint satisfaction

- Real-time computation

Advanced formulations include:

- Stochastic MPC for uncertainty
- Game-theoretic planning
- Learning-based cost functions

18.3.5 Multi-Agent Coordination

For N agents, the coupled optimal control problem:

$$\min \sum_i J_i[x_i, u_i, x_{-i}] \tag{18.28}$$

where x_{-i} are states of other agents.

Approaches:

- Centralized: Solve joint optimization
- Decentralized: Each agent solves own problem
- Distributed: Iterate to consensus

The action principle extends to:

- Formation control
- Collision avoidance
- Task allocation

Game-theoretic equilibria as stationary points of multi-agent action.

18.3.6 Learning and Adaptation

Combining optimal control with learning:

Adaptive optimal control:

- Update model online
- Dual control for exploration/exploitation

- Certainty equivalence
- Robust adaptive control

Iterative learning control: Improve over repetitions

$$J_{k+1} = J_k + \alpha \nabla_u J \tag{18.29}$$

Meta-learning: Learn to optimize

- MAML for few-shot adaptation
- Neural ODEs for dynamics
- Differentiable optimization layers

The action principle guides:

- What to optimize (cost function)
- How to optimize (algorithms)
- When to re-optimize (event triggers)

18.4 Trajectory Optimization

18.4.1 Direct Transcription Methods

Discretize the continuous problem:

$$\min \sum_{k=0}^{N-1} L(x_k, u_k)\Delta t + \phi(x_N) \tag{18.30}$$

subject to:

$$x_{k+1} = x_k + f(x_k, u_k)\Delta t \quad \text{(Euler)} \tag{18.31}$$

$$\text{or } x_{k+1} = x_k + \frac{\Delta t}{2}[f(x_k, u_k) + f(x_{k+1}, u_{k+1})] \quad \text{(Trapezoidal)} \tag{18.32}$$

This yields a nonlinear program (NLP):

- Variables: $\{x_0, ..., x_N, u_0, ..., u_{N-1}\}$

- Constraints: Dynamics, bounds, obstacles

Solve using:

- SQP (sequential quadratic programming)
- Interior point methods
- Augmented Lagrangian

18.4.2 Differential Dynamic Programming

For dynamics $\dot{x} = f(x, u)$, quadratize the cost-to-go:

$$V(x) \approx V_0 + V_x^T \delta x + \frac{1}{2}\delta x^T V_{xx} \delta x \tag{18.33}$$

Backward pass computes:

- Value function approximation
- Locally optimal feedback gains

Forward pass:

- Apply feedback policy
- Line search on step size

DDP exploits:

- Dynamic programming structure
- Local quadratic approximation
- Efficient recursion

18.4.3 Pseudospectral Methods

Approximate states and controls with polynomials:

$$x(t) \approx \sum_{i=0}^{N} x_i L_i(t) \tag{18.34}$$

where L_i are Lagrange polynomials at collocation points.

The action integral becomes:

$$S \approx \sum_k w_k L(x_k, u_k) \tag{18.35}$$

Dynamics enforced at collocation points:

$$\dot{X} = D \cdot X = F(X, U) \tag{18.36}$$

where D is differentiation matrix.
Advantages:

- Spectral accuracy
- Sparse NLP structure
- Natural handling of free final time

18.4.4 Convex Optimization Approaches

Convexify the problem for global optimality:
Linearization: Around reference trajectory

$$\min \sum [\delta x^T Q \delta x + \delta u^T R \delta u] \tag{18.37}$$

s.t. $\delta x_{k+1} = A\delta x_k + B\delta u_k$
Convex relaxation: For non-convex constraints

- SDP relaxation for quadratic constraints
- McCormick envelopes for bilinear terms
- Successive convex approximation

Change of variables: Transform to convex form

- Planetary landing: use log(mass)
- Robotics: inverse dynamics formulation

18.4.5 Trajectory Libraries and Motion Primitives

Precompute optimal trajectories offline:

Motion primitives: Short optimal segments

- Lattice-based planning
- Maneuver automaton
- Trajectory libraries

Basis functions: Parameterized trajectories

- Dynamic movement primitives (DMPs)
- Probabilistic movement primitives
- Neural trajectory generators

Online planning:

- Select and sequence primitives
- Optimize transitions
- Adapt to current state

The action principle determines:

- Primitive generation
- Optimal sequencing
- Smooth connections

18.4.6 Real-Time Implementation

Making trajectory optimization fast enough for control:

Warm starting: Use previous solution

- Shift and append
- Predictor-corrector schemes
- Continuation methods

Early termination: Suboptimal but feasible

- Fixed iteration budget
- Anytime algorithms
- Constraint prioritization

Parallel computation:

- GPU acceleration
- Distributed optimization
- Pipeline parallelism

Specialized hardware:

- FPGA implementation
- Custom ASICs
- Neuromorphic chips

The future: Real-time optimal control at kHz rates.

Conclusion

As we reach the end of our comprehensive journey through the principle of least action, we find ourselves not at a terminus, but at a vantage point from which we can survey the remarkable unity of physics and glimpse the horizons of future discovery. This principle, which began as an observation about the behavior of light rays and evolved into one of the most profound organizational principles in all of physics, has guided us through classical mechanics, field theory, relativity, quantum mechanics, and beyond.

Conceptual Unity Across Different Branches of Physics

The principle of least action reveals itself as a golden thread woven through the fabric of physical reality, connecting phenomena that might otherwise appear disparate and unrelated. From Fermat's observation that light travels along paths of least time, to Hamilton's formulation of classical mechanics, to Feynman's path integral approach to quantum mechanics, we have witnessed how a single variational principle can manifest across vastly different scales and regimes of physics.

In classical mechanics, the principle provides an elegant alternative to Newton's force-based approach, revealing conservation laws through Noether's theorem and offering powerful methods for solving complex problems. The Lagrangian and Hamiltonian formalisms, both rooted in the principle of least action, have proven to be more than mere reformulations—they are windows into the deeper structure of physical law. The symmetries that leave the action invariant directly correspond to conserved quantities: time translation symmetry yields energy conservation, spatial translation symmetry gives momentum conservation, and rotational symmetry provides angular momentum conservation. This profound connection between symmetry and conservation represents one of the most beautiful insights in theoretical physics.

The transition from classical to quantum mechanics illuminates another facet of the principle's universality. While classical particles follow the single path that extremizes the action, quantum particles explore all possible paths, with each path contributing to the total amplitude weighted by the exponential of the classical action. This quantum mechanical generalization, pioneered by Feynman, demonstrates that the classical principle emerges naturally from quantum mechanics in the appropriate limit—when the action is large compared to Planck's constant. The principle of least action thus serves as a bridge between the classical and quantum worlds, showing how macroscopic determinism emerges from microscopic probability.

In field theory, whether classical or quantum, the principle of least action continues to provide the fundamental framework. Maxwell's equations, which govern electromagnetic phenomena, emerge as the Euler-Lagrange equations from an appropriate action. The entire edifice of the Standard Model of particle physics is built upon actions that respect certain gauge symmetries. Even in general relativity, Einstein's field equations follow from extremizing the Einstein-Hilbert action, showing that the curvature of spacetime itself obeys a variational principle.

This universality extends beyond fundamental physics into complex systems, thermodynamics, and even engineering applications. The principle of least action in thermodynamics connects to maximum entropy principles, while in optimal control theory, it provides methods for finding efficient trajectories in robotics and aerospace engineering. The mathematical structure underlying all these applications—the calculus of variations—represents a unified language for describing nature's tendency toward optimal configurations.

Future Perspectives and Open Questions

Despite its tremendous success, the principle of least action continues to raise profound questions and point toward new frontiers in physics. Several key areas of investigation promise to deepen our understanding and potentially revolutionize our view of physical reality.

In quantum gravity, one of the most pressing questions concerns the nature of the action at the Planck scale. String theory proposes that the action should include contributions from extended objects—strings and branes—rather than just point particles. The Polyakov action for strings and its generalizations have led to rich mathematical structures and potential unifications of all fundamental forces. However, the principle of least action in string theory raises new conceptual challenges: what does it mean to extremize an action over the space of all possible worldsheets? How do we handle the infinite number of degrees of freedom inherent in extended objects?

Loop quantum gravity takes a different approach, attempting to quantize spacetime itself while preserving the diffeomorphism invariance of general relativity. Here, the challenge is formulating an appropriate action principle for discrete quantum geometry. The spin foam models that have emerged from this program can be viewed as a kind of path integral over quantum geometries, but the connection to a classical action principle remains subtle and not fully understood.

The cosmological constant problem represents another frontier where the principle of least action may need revision or reinterpretation. The observed acceleration of the universe's expansion suggests a small positive cosmological constant, but quantum field theory predictions for the vacuum energy density exceed observations by many orders of magnitude. This discrepancy might indicate that our understanding of how quantum effects contribute to the gravitational action is fundamentally incomplete.

Emergent gravity theories propose that spacetime and gravity might not be fundamental but could emerge from more basic quantum information-theoretic principles. In these approaches, the Einstein-Hilbert action would arise as an effective description, much like how thermodynamics emerges from statistical mechanics. This raises the question: is the principle of least action itself emergent, or does it represent something more fundamental about the nature of physical law?

In condensed matter physics, the study of topological phases of matter has revealed systems where the action includes topological terms that are not captured by local field theories. These Chern-Simons terms and their generalizations suggest that the principle of least action in many-body quantum systems may have richer structures than previously appreciated. The connection between entanglement entropy and geometry, exemplified by the AdS/CFT correspondence, hints at deep relationships between quantum information and the action principle that are only beginning to be explored.

The application of machine learning and artificial intelligence to physics raises intriguing questions about the principle of least action. Can neural networks discover new action principles from data? Might there be a meta-principle that determines the form of the action for different physical systems? Some researchers have begun exploring "physics-informed neural networks" that incorporate variational principles directly into their architecture, suggesting new computational approaches to solving complex physical problems.

Philosophical and Fundamental Implications

The principle of least action touches upon some of the deepest questions in the philosophy of science. Its teleological appearance—nature seems to "know" in advance which path to take—has puzzled thinkers since its inception. While we now understand that this teleology is apparent rather than real, emerging from the mathematical structure of the variational principle, the question remains: why should nature be describable by such principles at all?

One perspective views the principle of least action as evidence of an underlying simplicity and elegance in natural law. The fact that such diverse phenomena can be unified under a single mathematical framework suggests that nature, at its core, follows principles of economy and optimization. This resonates with the long-standing belief among physicists that the fundamental laws of nature should be expressible in simple, beautiful mathematical forms.

Another view interprets the principle more pragmatically, as a powerful organizational tool that happens to work rather than a fundamental truth about reality. From this perspective, the success of the action principle reflects the mathematics we have developed to describe nature rather than nature itself. The principle works because it efficiently encodes the differential equations governing physical systems, not because nature "computes" an action and "chooses" to minimize it.

The relationship between the principle of least action and causality deserves special attention. While Newton's laws express local causality—forces cause accelerations at each instant—the action principle appears non-local in time, considering entire trajectories at once. This tension is resolved by recognizing that the Euler-Lagrange equations derived from the action are local differential equations. Nevertheless, the action formulation suggests a block universe view where past, present, and future exist simultaneously, raising questions about the nature of time and free will.

The principle also connects to discussions about reductionism versus emergence in physics. While the action for fundamental particles might be simple, the effective actions for complex systems can exhibit rich, emergent behaviors not obviously present in the microscopic description. This hierarchy of effective actions at different scales provides a framework for understanding how complex phenomena arise from simple laws without invoking strong emergence or downward causation.

Information-theoretic perspectives on physics have suggested deep connections between the principle of least action and information theory. Some researchers propose that the action might be related to the information required to specify a physical process, with nature following paths that minimize this information. This view connects to ideas about computational complexity in physics and the physical basis of computation itself.

The anthropic principle and fine-tuning arguments also intersect with considerations of the action principle. Why should the actions governing our universe have the specific forms they do, with coupling constants and mass parameters that allow for complex chemistry and life? While these questions remain open, the principle of least action provides a framework for exploring how different choices of action would lead to different physics and potentially different universes.

As we conclude this exploration, we recognize that the principle of least action represents more than a calculational tool or even a fundamental law of physics. It embodies a way of thinking about nature that emphasizes unity, elegance, and deep mathematical structure. From its historical origins in optics to its modern applications in quantum field theory and beyond, the principle has proven to be one of the most fruitful concepts in the physicist's toolkit.

The journey through this book has taken us from the mathematical foundations laid by Euler and Lagrange, through the classical applications in mechanics and field theory, into the quantum realm where paths interfere and amplitudes reign, and finally to the frontiers where new physics may emerge. At each stage, the principle of least action has served as both guide and touchstone, revealing connections and suggesting new directions for investigation.

As physics continues to evolve, facing challenges from dark matter and dark energy to quantum gravity and the interpretation of quantum mechanics, the principle of least action will undoubtedly continue to play a central role. Whether it remains in its current form or undergoes profound generalizations, its spirit—the search for unifying principles that reveal nature's deep simplicity—will endure.

For the student embarking on the study of physics, mastery of the principle of least action opens doors to virtually every area of the field. For the researcher, it provides a framework for formulating new theories and discovering new phenomena. For the philosopher, it offers a window into the nature of physical law and the relationship between mathematics and reality.

In closing, we return to the wonder that a principle initially conceived to describe the path of light rays could encompass the motion of galaxies, the behavior of subatomic particles, and perhaps even the structure of spacetime itself. This remarkable scope reminds us that in physics, as in all of science, simple principles can have profound consequences, and the quest to understand nature's fundamental workings continues to reward us with insights both practical and sublime. The principle of least action stands as a testament to the power of human thought to uncover the hidden harmonies of the universe, and as an invitation to future generations to push further into the unknown, guided by the conviction that nature's deepest truths are both simple and beautiful.

Appendix A

Review of Differential and Integral Calculus

This appendix provides a comprehensive review of the calculus techniques essential for understanding the variational principles discussed throughout this book. While readers are assumed to have encountered these concepts before, we present them here with particular emphasis on their applications to physics and the calculus of variations.

A.1 Functions of Several Variables

The action functionals we encounter in physics typically depend on multiple variables—spatial coordinates, time, and often field values at different points. Understanding how to handle functions of several variables is therefore crucial. For a function $f(x_1, x_2, ..., x_n)$, the partial derivative $\partial f/\partial x_i$ represents the rate of change of f with respect to x_i while holding all other variables constant. This concept extends naturally to higher-order derivatives, with mixed partial derivatives satisfying the important property that, for smooth functions, the order of differentiation does not matter: $\partial^2 f/\partial x_i \partial x_j = \partial^2 f/\partial x_j \partial x_i$.

The total differential of a function provides a linear approximation to the change in the function value: $df = \sum_i (\partial f/\partial x_i) dx_i$. This expression becomes particularly important when we consider variations of action functionals, where we need to understand how small changes in the path or field configuration affect the action. The chain rule for functions of several variables takes the form: if $z = f(x, y)$ where $x = x(t)$ and

$y = y(t)$, then $dz/dt = (\partial f/\partial x)(dx/dt) + (\partial f/\partial y)(dy/dt)$. This rule is essential for deriving the Euler-Lagrange equations from the principle of stationary action.

The gradient vector $\nabla f = (\partial f/\partial x_1, \partial f/\partial x_2, ..., \partial f/\partial x_n)$ points in the direction of steepest increase of the function and plays a crucial role in understanding extrema of functions. For a function to have an extremum at a point, its gradient must vanish there—a condition that generalizes to the vanishing of the functional derivative for action functionals. The Hessian matrix, containing all second partial derivatives, determines whether an extremum is a minimum, maximum, or saddle point, analogous to how the second variation of an action determines the stability of a stationary path.

A.2 Taylor Series and Approximations

Taylor series expansions provide powerful tools for approximating functions and understanding their behavior near specific points. For a function of one variable, the Taylor series about a point a is: $f(x) = f(a) + f'(a)(x-a) + f''(a)(x-a)^2/2! + ...$ This expansion extends to functions of several variables through multi-index notation, becoming particularly useful when analyzing small deviations from equilibrium configurations or when developing perturbation theory.

In physics applications, we often truncate Taylor series to obtain tractable approximations. The harmonic oscillator, fundamental to many physical systems, arises from keeping only the quadratic term in the potential energy expansion about a stable equilibrium. Higher-order terms lead to anharmonic corrections that can be treated perturbatively. The validity of such approximations depends on the convergence properties of the series and the magnitude of the expansion parameter.

Asymptotic expansions, which may not converge but provide excellent approximations for large or small parameter values, complement Taylor series in physics. These become essential when dealing with WKB approximations in quantum mechanics or analyzing the classical limit where $\hbar \to 0$. Understanding when and how to use different types of series expansions is crucial for extracting physical insights from mathematical expressions.

A.3 Integration Techniques

Integration appears throughout physics in calculating actions, partition functions, and expectation values. Beyond basic integration formulas, several techniques prove particularly valuable. Integration by parts, $\int u\,dv = uv - \int v\,du$, is essential in deriving the Euler-Lagrange equations and appears repeatedly in the calculus of variations. When applied to definite integrals with appropriate boundary conditions, surface terms often vanish, simplifying many calculations.

Change of variables in multiple integrals requires careful attention to the Jacobian determinant. For a transformation from coordinates $(x_1, ..., x_n)$ to $(y_1, ..., y_n)$, the volume element transforms as $dx_1...dx_n = |\partial(x_1, ..., x_n)/\partial(y_1, ..., y_n)|dy_1...dy_n$. This transformation law becomes crucial when changing between different coordinate systems in mechanics or when performing canonical transformations in phase space. In field theory, functional integrals generalize these concepts to infinite-dimensional spaces.

Contour integration in the complex plane provides powerful techniques for evaluating real integrals and understanding analytic properties of physical quantities. Residue theory allows us to evaluate integrals by considering singularities of the integrand, while saddle point methods approximate integrals dominated by contributions near stationary points of the phase. These techniques become essential in quantum field theory calculations and statistical mechanics, where Gaussian integrals and their generalizations appear frequently.

A.4 Differential Equations

The Euler-Lagrange equations derived from the principle of least action are differential equations, making their study central to our subject. Linear differential equations with constant coefficients can be solved systematically using characteristic equations, yielding solutions involving exponentials and trigonometric functions. These solutions describe small oscillations about equilibrium in many physical systems.

For nonlinear differential equations, which arise from more realistic physical models, analytical solutions rarely exist. However, qualitative methods from dynamical systems theory—including phase plane analysis, stability theory, and bifurcation analysis—provide crucial insights.

Fixed points and their stability determine the long-term behavior of systems, while phase portraits visualize the global dynamics. These tools become particularly important when studying integrable and chaotic Hamiltonian systems.

Partial differential equations (PDEs) govern field theories and continuous media. The wave equation, diffusion equation, and Laplace's equation represent fundamental types that appear throughout physics. Solution methods include separation of variables, Green's functions, and transform techniques. The method of characteristics proves particularly valuable for first-order PDEs and connects to the Hamilton-Jacobi equation in mechanics. Understanding the classification of PDEs—elliptic, parabolic, or hyperbolic—helps determine appropriate solution strategies and physical interpretations.

A.5 Functional Analysis Concepts

Since actions are functionals—mappings from functions to numbers—some concepts from functional analysis prove invaluable. A functional $S[f]$ assigns a number to each function f from some appropriate space. The functional derivative $\delta S/\delta f(x)$ generalizes the partial derivative to this infinite-dimensional setting and satisfies: $S[f + \varepsilon\eta] = S[f] + \varepsilon \int (\delta S/\delta f(x))\eta(x)dx + O(\varepsilon^2)$, where η is an arbitrary test function vanishing at boundaries.

Function spaces provide the rigorous setting for variational problems. The space $C^1[a, b]$ of continuously differentiable functions on $[a, b]$ often serves as the domain for classical mechanical actions. For field theories, Sobolev spaces incorporate derivatives in their norm definitions, ensuring sufficient regularity for physical applications. The choice of function space can affect whether minimizers exist and whether they are unique.

Operators on function spaces, particularly linear operators, play crucial roles in physics. The Sturm-Liouville theory for self-adjoint operators provides the mathematical foundation for quantum mechanics and explains the discreteness of energy levels in bound systems. Spectral theory connects eigenvalues of operators to physical observables, while perturbation theory for operators underlies many approximation schemes in physics.

Appendix B

Introduction to Lie Groups and Algebras

Lie groups and their associated Lie algebras provide the mathematical framework for understanding continuous symmetries in physics. From the rotation group in classical mechanics to gauge groups in particle physics, these structures appear throughout modern physics. This appendix introduces the key concepts and develops the tools needed for applications to physical theories.

B.1 Lie Groups: Definitions and Examples

A Lie group is simultaneously a group and a smooth manifold, with the group operations (multiplication and inversion) being smooth maps. This combination of algebraic and geometric structures makes Lie groups the natural setting for continuous symmetries. The dimension of a Lie group as a manifold corresponds to the number of continuous parameters needed to specify group elements.

Physical examples abound. The rotation group SO(3) consists of 3×3 orthogonal matrices with determinant 1, describing rotations in three-dimensional space. As a manifold, SO(3) is three-dimensional, with elements parameterized by three angles (though no single parameterization covers the entire group without singularities). The special unitary group SU(2), describing spin-1/2 rotations in quantum mechanics, is also three-dimensional but has a simpler topology—it is homeomorphic to the three-sphere S^3.

The general linear group $\mathrm{GL}(n, \mathbb{R})$ of invertible $n \times n$ real matrices is an n^2-dimensional Lie group that contains many important subgroups. The orthogonal group $\mathrm{O}(n)$ preserves inner products, while its subgroup $\mathrm{SO}(n)$ additionally preserves orientation. In special relativity, the Lorentz group O(3,1) preserves the spacetime interval, with the proper orthochronous subgroup $\mathrm{SO}^+(3,1)$ preserving both spatial orientation and time direction. Complex Lie groups like $\mathrm{U}(n)$ and $\mathrm{SU}(n)$ become essential in quantum mechanics and gauge theories.

Matrix Lie groups, while not exhausting all possibilities, suffice for most physics applications. Every compact Lie group is isomorphic to a matrix group, and even non-compact groups relevant to physics typically have faithful matrix representations. This allows us to leverage linear algebra in studying their properties.

B.2 Lie Algebras: The Infinitesimal Structure

The Lie algebra of a Lie group captures its infinitesimal structure—the behavior near the identity element. For matrix groups, Lie algebra elements are matrices X such that $\exp(tX)$ lies in the group for all real t. The exponential map $\exp : \mathfrak{g} \to G$ connects the Lie algebra $\mathfrak{g}$ to the Lie group G, with $\exp(X) = I + X + X^2/2! + ...$ providing the explicit formula for matrix groups.

The Lie bracket $[X, Y] = XY - YX$ endows the Lie algebra with additional structure beyond that of a vector space. This bracket operation is bilinear, antisymmetric, and satisfies the Jacobi identity $[[X, Y], Z] + [[Y, Z], X] + [[Z, X], Y] = 0$. Physically, the Lie bracket corresponds to the commutator of infinitesimal transformations, determining how symmetry operations fail to commute.

Structure constants f_{ij}^k defined by $[T_i, T_j] = \sum_k f_{ij}^k T_k$ for a basis $\{T_i\}$ of the Lie algebra encode all information about the bracket operation. These constants are antisymmetric in the lower indices and satisfy constraints from the Jacobi identity. For compact simple Lie groups, the structure constants can be chosen to be completely antisymmetric in all three indices.

The Lie algebra homomorphism property—that representations of the group induce representations of the algebra with the bracket going to the commutator—makes Lie algebras powerful computational tools. Instead of dealing with the full nonlinear structure of the group, we can work with the linear structure of its algebra. This simplification proves especially valuable in quantum mechanics, where observables correspond to Lie algebra elements.

B.3 Representations and Physics

A representation of a Lie group G on a vector space V is a smooth homomorphism $\rho : G \to \mathrm{GL}(V)$, assigning to each group element a linear transformation of V that preserves the group structure. In physics, representation spaces correspond to spaces of states, with group elements acting as symmetry transformations. The trivial representation maps every element to the identity, while the defining representation of a matrix group is the natural action on column vectors.

Irreducible representations (irreps) cannot be decomposed into smaller representations and play a fundamental role in physics. States transforming under an irrep of the symmetry group cannot be further separated by symmetry considerations alone. For compact Lie groups, every representation decomposes as a direct sum of irreps, with each irrep appearing with some multiplicity. This decomposition corresponds physically to identifying the symmetry content of a system.

The representation theory of SU(2) illustrates key concepts. Its irreps are labeled by spin $j = 0, 1/2, 1, 3/2, ...$, with dimension $2j + 1$. The spin-1/2 representation is the defining representation on $\mathbb{C}^2$, while higher spins arise as symmetric tensor powers. The Clebsch-Gordan series describes how tensor products decompose: $j_1 \otimes j_2 = |j_1 - j_2| \oplus ... \oplus (j_1 + j_2)$, providing rules for combining angular momenta in quantum mechanics.

For non-compact groups like the Lorentz group, representation theory becomes more subtle. Unitary irreps may be infinite-dimensional, corresponding to the continuous spectrum of momenta in relativistic quantum mechanics. The classification of these representations by Wigner connects to the possible types of elementary particles, with mass and spin labeling the irreps of the Poincaré group.

B.4 Root Systems and Classification

The structure of a Lie algebra is elegantly encoded in its root system, particularly for semisimple Lie algebras. The Cartan subalgebra $\mathfrak{h}$ consists of mutually commuting elements, with its dimension called the rank of the algebra. In the adjoint representation, Cartan subalgebra elements act diagonalizably, with eigenvalues $\alpha(H)$ for $H \in \mathfrak{h}$ called roots. The corresponding eigenspaces are one-dimensional for non-zero roots.

Root vectors E_α satisfy $[H, E_\alpha] = \alpha(H)E_\alpha$ for all $H \in \mathfrak{h}$, spanning root spaces that organize the Lie algebra. The commutation relations $[E_\alpha, E_\beta] = N_{\alpha\beta}E_{\alpha+\beta}$ (with $N_{\alpha\beta} = 0$ if $\alpha+\beta$ is not a root) show how roots add under commutation. This structure generalizes the familiar raising and lowering operators of angular momentum to arbitrary semisimple Lie algebras.

Simple roots form a basis for the root system with the property that all positive roots are non-negative integer linear combinations of simple roots. The Dynkin diagram encodes the angles between simple roots, providing a complete classification of simple Lie algebras: the classical series A_n (corresponding to $\mathrm{SU}(n+1)$), B_n ($\mathrm{SO}(2n+1)$), C_n ($\mathrm{Sp}(2n)$), D_n ($\mathrm{SO}(2n)$), and the exceptional algebras E_6, E_7, E_8, F_4, and G_2.

Weights generalize roots to arbitrary representations. A weight vector v_λ satisfies $\rho(H)v_\lambda = \lambda(H)v_\lambda$ for all H in the Cartan subalgebra, with λ the weight. Highest weight vectors are annihilated by all positive root vectors and uniquely determine irreducible representations. This highest weight theory provides a systematic way to construct and classify all representations of semisimple Lie algebras.

B.5 Applications to Gauge Theories

In gauge theories, Lie groups appear as local symmetries, with gauge fields taking values in the Lie algebra. For a gauge group G, the gauge potential $A_\mu = \sum_a A_\mu^a \tau_a$ expands in terms of Lie algebra generators τ_a. Under gauge transformations $g(x) \in G$, the potential transforms as $A'_\mu = gA_\mu g^{-1} - (\partial_\mu g)g^{-1}$, ensuring gauge covariance of physical quantities.

The field strength tensor $F_{\mu\nu} = \partial_\mu A_\nu - \partial_\nu A_\mu + [A_\mu, A_\nu]$ generalizes the electromagnetic field tensor to non-Abelian gauge theories. Its transformation $F'_{\mu\nu} = gF_{\mu\nu}g^{-1}$ is simpler than that of the gauge potential, making it the natural object for constructing gauge-invariant actions. The Yang-Mills action $S = -(1/4)\int \mathrm{Tr}(F_{\mu\nu}F^{\mu\nu})d^4x$ provides the kinetic term for gauge fields.

Matter fields transform in representations of the gauge group, with covariant derivatives $D_\mu\psi = \partial_\mu\psi + A_\mu\psi$ ensuring gauge-invariant kinetic terms. The specific representation determines the matter field's interactions with gauge bosons. In the Standard Model, quarks transform in the fundamental representation of $\mathrm{SU}(3)_{\mathrm{color}}$, while leptons are singlets. The electroweak $\mathrm{SU}(2)_L\times \mathrm{U}(1)_Y$ structure assigns left-handed fermions to doublets and right-handed fermions to singlets.

Spontaneous symmetry breaking, implemented through the Higgs mechanism, reduces the gauge group to a subgroup. Gauge bosons corresponding to broken generators acquire mass, while those for unbroken generators remain massless. This mechanism requires scalar fields in appropriate representations, with non-zero vacuum expectation values breaking the symmetry while preserving gauge invariance of the Lagrangian.

Appendix C

Asymptotic Methods

Asymptotic methods provide powerful techniques for approximating solutions to problems involving small or large parameters. These methods often yield excellent approximations even when exact solutions are intractable, making them indispensable tools in theoretical physics. This appendix develops the key techniques and illustrates their applications to physical problems.

C.1 Asymptotic Expansions and Series

An asymptotic expansion of a function $f(x)$ as $x \to x_0$ (often $x_0 = 0$ or ∞) is a series $\sum_n a_n \phi_n(x)$ such that for each N, the remainder $R_N(x) = f(x) - \sum_{n=0}^{N} a_n \phi_n(x)$ satisfies $R_N(x) = o(\phi_N(x))$ as $x \to x_0$. Unlike convergent series, asymptotic series may diverge, but truncated series provide increasingly accurate approximations as x approaches x_0. The optimal truncation often occurs near the smallest term.

The gauge functions $\phi_n(x)$ form an asymptotic sequence: $\phi_{n+1}(x) = o(\phi_n(x))$ as $x \to x_0$. Common examples include powers x^n, exponentials e^{-x}, and logarithms $\log^n(x)$. For oscillatory problems, gauge functions may include trigonometric terms like $\sin(x^n)$ or $\cos(x^n)$. The choice of gauge functions depends on the problem's structure and the dominant behavior near x_0.

Stirling's approximation $n! \sim \sqrt{2\pi n}(n/e)^n e^{1/12n}$ exemplifies an asymptotic expansion, with the factor $e^{1/12n}$ beginning an asymptotic series in powers of $1/n$. This approximation proves invaluable in statistical mechanics for evaluating partition functions and entropy in the thermodynamic limit. The error in truncating after the first term is less than 1% for $n > 10$, demonstrating the practical utility of asymptotic approximations.

Operations on asymptotic expansions—addition, multiplication, differentiation, and sometimes integration—follow specific rules. Term-by-term differentiation is generally valid, but integration requires care due to potential small exponential terms that differentiation annihilates but integration amplifies. These subtleties become crucial when deriving asymptotic solutions to differential equations.

C.2 Method of Steepest Descent

Integrals of the form $I(\lambda) = \int_C f(z)e^{\lambda g(z)}dz$ for large λ arise throughout physics, from partition functions to wave propagation. The method of steepest descent (saddle point method) evaluates such integrals by deforming the contour C to pass through saddle points of $g(z)$ along paths of steepest descent, where the integrand's oscillations are minimized.

At a saddle point z_0 where $g'(z_0) = 0$, the local behavior $g(z) \approx g(z_0) + (1/2)g''(z_0)(z - z_0)^2$ determines the leading contribution. Choosing the contour such that $g''(z_0)(z - z_0)^2$ is negative real ensures exponential decay away from the saddle. The leading-order result $I(\lambda) \sim \sqrt{2\pi/\lambda|g''(z_0)|}f(z_0)e^{\lambda g(z_0)}$ resembles a Gaussian integral, with corrections forming an asymptotic series in powers of $1/\lambda$.

Multiple saddle points may contribute, with their relative importance determined by $\mathrm{Re}[g(z_i)]$. In the limit $\lambda \to \infty$, only saddles with maximal $\mathrm{Re}[g(z)]$ contribute significantly. Phase transitions can occur when two saddles exchange dominance, leading to non-analytic behavior in physical quantities. This mechanism underlies many phase transitions in statistical mechanics and quantum field theory.

The method extends to multidimensional integrals through the stationary phase approximation. For rapidly oscillating integrands, significant contributions come only from regions where the phase is stationary. This principle explains classical limits in quantum mechanics: as $\hbar \to 0$, path integrals are dominated by classical paths where the action is stationary.

C.3 WKB Approximation

The Wentzel-Kramers-Brillouin (WKB) method provides asymptotic solutions to differential equations with slowly varying coefficients. For the Schrödinger equation $-\hbar^2\psi''/2m + V(x)\psi = E\psi$, the WKB ansatz $\psi(x) = A(x)\exp(iS(x)/\hbar)$ leads to a series expansion for $S(x)$ in powers of $\hbar$. The leading term S_0 satisfies the classical Hamilton-Jacobi equation $(dS_0/dx)^2/2m + V(x) = E$, connecting quantum and classical mechanics.

In classically allowed regions where $E > V(x)$, the WKB solution takes the form $\psi(x) \sim C/\sqrt{p(x)}\exp\left(\pm i\int p(x)dx/\hbar\right)$, with $p(x) = \sqrt{2m(E - V(x))}$ the classical momentum. The $1/\sqrt{p}$ factor ensures current conservation. In classically forbidden regions where $E < V(x)$, exponentially growing and decaying solutions replace the oscillatory behavior.

Connection formulas relate WKB solutions across turning points where $E = V(x)$. Near a simple turning point, the potential is approximately linear, and the exact solution involves Airy functions. Matching asymptotic forms of Airy functions to WKB solutions on either side yields connection formulas essential for determining bound state energies and tunneling probabilities.

The Bohr-Sommerfeld quantization condition $\oint p\,dx = 2\pi\hbar(n + \mu)$ emerges from requiring single-valuedness of the wave function, with $\mu = 1/2$ for soft turning points. This semiclassical quantization reproduces exact energy levels for harmonic oscillators and gives excellent approximations for many potentials when quantum numbers are large. Extensions to multidimensional systems involve Maslov indices accounting for caustics where the classical action becomes multivalued.

C.4 Multiple Scale Analysis

Physical systems often involve processes occurring on disparate time or length scales. Multiple scale analysis systematically accounts for these different scales by introducing multiple independent variables. For a problem with a small parameter ε, we might introduce fast time $\tau = t$ and slow time $T = \varepsilon t$, treating them as independent in the perturbation expansion.

Consider a weakly nonlinear oscillator: $\ddot{x} + x + \varepsilon f(x, \dot{x}) = 0$. Naive perturbation theory yields secular terms growing with time, invalidating the expansion. Multiple scale analysis with $x(t, \varepsilon) = X_0(\tau, T) + \varepsilon X_1(\tau, T) + \ldots$ and the chain rule $d/dt = \partial/\partial\tau + \varepsilon\partial/\partial T$ eliminates secular terms by imposing solvability conditions. These conditions determine the slow evolution of amplitude and phase.

The method reveals how nonlinearities cause frequency shifts and amplitude-dependent effects. For the Duffing oscillator with $f(x, \dot{x}) = x^3$, the frequency shifts as $\omega = 1 + 3\varepsilon A^2/8 + O(\varepsilon^2)$, where A is the slowly varying amplitude. This frequency-amplitude relationship explains many phenomena in nonlinear dynamics, from musical instruments to particle accelerators.

Applications extend to partial differential equations, where multiple spatial or temporal scales arise naturally. Wave modulation in dispersive media, boundary layer theory in fluid dynamics, and homogenization of composite materials all employ multiple scale techniques. The method's systematic nature allows computer algebra systems to automate calculations, making it practical for complex problems.

C.5 Matched Asymptotic Expansions

When different approximations hold in different regions, matched asymptotic expansions connect these regional solutions to form a global approximation. This technique proves essential for boundary layer problems, where rapid variations near boundaries transition to slowly varying bulk behavior.

The general procedure involves: (1) identifying regions with different characteristic scales, (2) introducing appropriate scaled variables in each region, (3) developing asymptotic expansions in each region, and (4) matching expansions in overlap regions where both approximations are valid. The matching principle states that the outer limit of the inner expansion equals the inner limit of the outer expansion.

Boundary layer theory in fluid dynamics exemplifies the method. For high Reynolds number flow past a body, viscous effects are negligible except in thin boundary layers. The outer inviscid solution satisfies Euler equations but cannot satisfy no-slip boundary conditions. The inner boundary layer solution, using stretched coordinates, includes viscous effects and matches to the outer solution away from the surface.

Quantum mechanical tunneling through barriers also employs matched asymptotics. WKB solutions in classically allowed regions must be matched through the barrier region where different approximations apply. The matching determines tunneling probabilities crucial for understanding radioactive decay, field emission, and scanning tunneling microscopy. Extensions to multidimensional tunneling involve complex trajectories and instanton methods.

Appendix D

Integrable Hamiltonian Systems and Chaos

The study of Hamiltonian systems reveals a rich interplay between regular, integrable motion and chaotic dynamics. This appendix explores both extremes and the transitions between them, providing tools for analyzing the dynamical systems arising from the principle of least action.

D.1 Integrable Systems

A Hamiltonian system with n degrees of freedom is completely integrable if it possesses n independent conserved quantities (first integrals) in involution: $\{I_i, I_j\} = 0$ for all i, j, where $\{,\}$ denotes the Poisson bracket. The Hamiltonian H is typically one of these integrals. The Arnold-Liouville theorem states that if the level sets of the integrals are compact and connected, they are diffeomorphic to n-dimensional tori, with motion on these tori being quasiperiodic.

Action-angle variables (I, θ) provide the natural coordinates for integrable systems. The actions $I_i = (1/2\pi) \oint p_i dq_i$ are adiabatic invariants, remaining constant under slow parameter changes. The angles θ_i increase linearly with time: $\dot{\theta}_i = \omega_i(I) = \partial H/\partial I_i$, with frequencies depending only on the actions. The motion is periodic if frequency ratios are rational, quasiperiodic if irrational.

Classical examples include the Kepler problem, where the Runge-Lenz vector provides additional integrals beyond energy and angular momentum. The harmonic oscillator, free particle, and rigid body are also integrable. Separable systems, where the Hamilton-Jacobi equation admits separated solutions, form a large class of integrable systems. Separation is possible in specific coordinate systems determined by the system's symmetries.

The inverse scattering transform provides a powerful method for solving certain nonlinear partial differential equations, revealing their integrable structure. The Korteweg-de Vries equation, nonlinear Schrödinger equation, and sine-Gordon equation are integrable in this sense. Soliton solutions, which maintain their shape despite nonlinearity, emerge naturally from the inverse scattering formalism. The existence of infinitely many conserved quantities characterizes these infinite-dimensional integrable systems.

D.2 Canonical Perturbation Theory

Real physical systems rarely exhibit perfect integrability, but many are close to integrable systems. Canonical perturbation theory systematically studies such nearly integrable systems, with the Hamiltonian $H = H_0(I) + \varepsilon H_1(I, \theta, t)$, where H_0 is integrable and ε is small. The goal is to find a canonical transformation to new variables where the motion is simpler.

Classical perturbation theory seeks transformations eliminating angle dependence to some order in ε. The generating function $F(I', \theta) = I' \cdot \theta + \varepsilon S_1(I', \theta) + \varepsilon^2 S_2(I', \theta) + ...$ produces new actions I' that are approximately conserved. However, small denominators $\omega_i(I') \cdot k$ for integer vectors k near resonances complicate the procedure. These denominators can vanish, causing the perturbation series to diverge.

The Kolmogorov-Arnold-Moser (KAM) theorem addresses this difficulty, proving that most invariant tori of integrable systems persist under small perturbations, albeit with slight deformations. Tori with sufficiently irrational frequency ratios (satisfying Diophantine conditions $|\omega \cdot k| \geq \gamma/|k|^\tau$ for some $\gamma > 0$ and $\tau > n - 1$) survive. The measure of destroyed tori vanishes as $\varepsilon \to 0$, but their destruction can have profound dynamical consequences.

Resonant tori, where frequency ratios are rational, are generally destroyed by perturbations. Near resonances, the dynamics is approximately described by a pendulum-like Hamiltonian, with stable and unstable periodic orbits replacing the original torus. Chains of islands form around stable orbits, with chaotic layers near separatrices. Secondary resonances within islands create self-similar structures observable in Poincaré sections.

D.3 Onset of Chaos

As perturbation strength increases, more tori are destroyed, and chaotic regions grow. The last KAM torus to be destroyed typically has the golden mean frequency ratio, being "most irrational." Its destruction often marks a transition to global chaos, where trajectories can explore large regions of phase space.

Homoclinic tangles, formed when stable and unstable manifolds of hyperbolic points intersect transversely, create complicated dynamics. The horseshoe map paradigm explains how stretching and folding in phase space generates sensitive dependence on initial conditions. Trajectories near homoclinic intersections exhibit chaotic behavior, with positive Lyapunov exponents measuring the exponential divergence of nearby trajectories.

The Chirikov overlap criterion estimates when large-scale chaos emerges. When resonance islands grow large enough to overlap, invariant tori between them are destroyed, allowing diffusion through phase space. For the standard map, this occurs around $K = 0.97$, providing a concrete example of the transition from regular to chaotic motion.

Chaos in Hamiltonian systems differs qualitatively from dissipative chaos. Phase space volume preservation (Liouville's theorem) prevents attractors, and chaotic regions coexist with regular islands at all scales. This mixed phase space structure has important implications for transport and statistical properties. Cantori (remnants of destroyed tori) can significantly impede transport, creating stickiness near regular regions.

D.4 Chaos Detection and Characterization

Several methods exist for detecting and quantifying chaos in Hamiltonian systems. Lyapunov exponents measure the average exponential rate of divergence of nearby trajectories. For Hamiltonian systems, Lyapunov exponents come in pairs $\pm\lambda$, with positive exponents indicating chaos. The Kolmogorov-Sinai entropy $h = \sum_{\lambda_i>0} \lambda_i$ quantifies the rate of information production.

Poincaré sections reduce the continuous flow to a discrete map by recording intersections with a codimension-one surface. Regular orbits appear as curves (invariant tori) or points (periodic orbits), while chaotic orbits fill areas densely. The fractal dimension of chaotic regions provides another characterization, with box-counting or correlation dimensions commonly used.

Frequency analysis offers a powerful alternative to traditional methods. The fundamental frequencies of quasiperiodic motion appear as sharp peaks in power spectra, while chaotic motion produces broadband spectra. Numerical analysis of fundamental frequencies can distinguish between regular and chaotic motion with high precision, even for weakly chaotic orbits where other methods fail.

The 0-1 test for chaos provides a binary diagnostic using bounded time series. By examining the growth rate of certain statistics, it distinguishes between regular (0) and chaotic (1) dynamics without requiring phase space reconstruction or Lyapunov exponent calculation. This makes it particularly useful for experimental data where the underlying equations are unknown.

D.5 Applications and Modern Developments

Understanding integrable systems and chaos has profound implications across physics. In celestial mechanics, KAM theory explains the stability of planetary orbits despite perturbations from other planets. The asteroid belt's structure, with gaps at resonances with Jupiter, exemplifies the interplay between regular and chaotic dynamics.

Quantum chaos studies how classical chaos manifests in quantum systems. While quantum evolution is linear and deterministic, signatures of classical chaos appear in energy level statistics, eigenfunction morphology, and semiclassical approximations. Random matrix theory successfully describes statistical properties of chaotic quantum systems, with different symmetry classes corresponding to different random matrix ensembles.

Plasma physics and accelerator physics heavily utilize Hamiltonian dynamics. Particle confinement in magnetic fields, beam stability in accelerators, and transport in fusion devices all involve understanding regular and chaotic motion in high-dimensional phase spaces. Modern computational methods combine analytical estimates with symplectic integration schemes that preserve the Hamiltonian structure.

Recent developments include the study of Arnold diffusion in higher-dimensional systems, where chaos can cause slow diffusion along resonance channels even when KAM tori confine motion in lower dimensions. Nekhoroshev stability provides exponentially long stability times for nearly integrable systems, important for long-term stability questions. The anti-integrable limit, where perturbation dominates, offers another approach to understanding complex dynamics, with chaotic motion emerging from the continuation of uncoupled states.

Bibliography

Historical Sources and Classical Foundations

Fermat, Pierre de (1662). "Synthesis ad refractiones." In *Œuvres de Fermat*, Vol. 1, edited by Charles Henry and Paul Tannery. Paris: Gauthier-Villars, 1891.

This seminal work introduces Fermat's principle of least time, marking the first variational principle in physics. Fermat demonstrated that light travels between two points along the path requiring the least time, successfully explaining the laws of reflection and refraction. While the original manuscript was lost, this reconstruction from Fermat's correspondence reveals his geometric approach and philosophical insights that would profoundly influence the development of analytical mechanics. The principle's teleological appearance—that light somehow "knows" which path to take—sparked centuries of debate about causality and the nature of physical laws.

Maupertuis, Pierre Louis Moreau de (1744). "Accord de différentes lois de la nature qui avaient jusqu'ici paru incompatibles." *Mémoires de l'Académie Royale des Sciences de Paris*, pp. 417-426.

Maupertuis extended variational principles from optics to mechanics, proposing that nature acts to minimize "action," which he defined somewhat vaguely as the product of mass, velocity, and distance. Despite its mathematical imprecision, this work established the philosophical foundation for least action principles in mechanics. Maupertuis's emphasis on economy and simplicity in nature influenced subsequent developments by Euler and Lagrange. The paper also ignited priority disputes with Leibniz's followers and philosophical controversies about divine design in nature, documented in the extensive correspondence of the period.

Euler, Leonhard (1744). *Methodus inveniendi lineas curvas maximi minimive proprietate gaudentes.* Lausanne: Bousquet.

Euler's masterpiece on the calculus of variations provided the mathematical framework necessary for rigorous formulation of variational principles. This work introduced the differential equation now known as the Euler-Lagrange equation and demonstrated its application to numerous problems in geometry and mechanics. Euler showed how to find curves that maximize or minimize integral quantities, establishing techniques still fundamental to modern physics. His appendix on elastic curves and his treatment of isoperimetric problems laid groundwork for constrained optimization in physics.

Lagrange, Joseph-Louis (1788). *Mécanique analytique.* Paris: Desaint.

This revolutionary treatise transformed mechanics from a geometric to an analytical science, presenting all of mechanics through the principle of virtual work and variational methods. Lagrange proudly noted that his book contained no diagrams, emphasizing the power of analytical methods. He introduced generalized coordinates, derived what we now call Lagrange's equations, and showed how constraints could be incorporated through his multiplier method. The work's systematic approach and elegant formalism established the template for theoretical physics presentations that persists today. Modern editions with extensive commentary help readers appreciate Lagrange's innovations in context.

Hamilton, William Rowan (1834-1835). "On a General Method in Dynamics." *Philosophical Transactions of the Royal Society*, Parts I and II.

Hamilton's papers introduced his principle of varying action and the canonical formalism that bears his name. By focusing on the action as a function of endpoints rather than paths, Hamilton discovered the profound connection between mechanics and optics, showing that particle trajectories and light rays obey analogous principles. His introduction of the Hamiltonian function and canonical equations provided a formalism particularly suited to perturbation theory and later quantum mechanics. These papers also developed the Hamilton-Jacobi equation, connecting mechanics to partial differential equations and presaging the wave nature of matter.

Modern Classical Mechanics Texts

Goldstein, Herbert, Poole, Charles P., and Safko, John L. (2002). *Classical Mechanics*, 3rd Edition. San Francisco: Addison-Wesley.

The standard graduate-level text for classical mechanics, Goldstein provides comprehensive coverage of Lagrangian and Hamiltonian mechanics with extensive historical notes. The book excels in its treatment of canonical transformations, Hamilton-Jacobi theory, and the connection to quantum mechanics. Particularly valuable are the chapters on rigid body dynamics and small oscillations, which demonstrate the power of Lagrangian methods. The third edition updates the mathematical presentation and includes new sections on chaos and nonlinear dynamics, though some prefer the cleaner exposition of earlier editions.

Landau, L.D. and Lifshitz, E.M. (1976). *Mechanics*, 3rd Edition. Oxford: Butterworth-Heinemann.

This concise yet profound text exemplifies the Russian approach to theoretical physics: elegant, efficient, and demanding. Starting from the principle of least action, Landau and Lifshitz derive all of classical mechanics with remarkable economy. Their treatment of symmetries and conservation laws via Noether's theorem is particularly illuminating. The problems, while challenging, guide readers to deep insights. The book's discussion of mechanical similarity and scaling arguments provides tools used throughout theoretical physics. Despite its brevity, or perhaps because of it, this remains one of the most influential mechanics texts.

Arnold, V.I. (1989). *Mathematical Methods of Classical Mechanics*, 2nd Edition. New York: Springer-Verlag.

Arnold's text approaches classical mechanics from a rigorous mathematical perspective, emphasizing the geometric structures underlying physical theories. The treatment of symplectic geometry, completely integrable systems, and perturbation theory is unmatched. Arnold introduces modern mathematical concepts naturally, showing why abstract mathematics illuminates physics. The book's appendices on Riemannian geometry, Lie groups, and contact structures provide gateways to advanced topics. While mathematically demanding, Arnold rewards careful study with deep understanding of the geometric nature of mechanics.

José, Jorge V. and Saletan, Eugene J. (1998). *Classical Dynamics: A Contemporary Approach.* Cambridge: Cambridge University Press.

This modern text balances traditional approaches with contemporary developments in nonlinear dynamics and chaos. The authors present Lagrangian and Hamiltonian mechanics in parallel, highlighting their complementary insights. Extensive numerical examples and computational exercises reflect modern practice in theoretical physics. The book's treatment of nonholonomic constraints, time-dependent transformations, and geometric phases addresses topics often omitted from standard texts. Clear prose and careful attention to mathematical details make this an excellent bridge between undergraduate and graduate study.

Quantum Mechanics and Path Integrals

Feynman, R.P. and Hibbs, A.R. (1965). *Quantum Mechanics and Path Integrals.* New York: McGraw-Hill. (Emended edition by Daniel F. Styer, 2010, Dover Publications)

Feynman's path integral formulation of quantum mechanics represents one of the most profound reinterpretations of a physical theory. This book, developed from Feynman's original notes with Hibbs, shows how quantum mechanics emerges from a sum over all possible paths, each weighted by the exponential of the classical action. The approach provides intuitive understanding of quantum phenomena and powerful calculational tools. The emended edition corrects numerous errors in the original while preserving Feynman's unique pedagogical style. Applications to statistical mechanics demonstrate the formalism's versatility beyond quantum mechanics.

Schulman, L.S. (1981). *Techniques and Applications of Path Integration.* New York: Wiley-Interscience. (Reprinted by Dover, 2005)

Schulman's comprehensive treatment of path integration techniques has become the standard reference for practical calculations. The book develops approximation methods, discusses convergence issues rigorously, and provides extensive applications from quantum mechanics to polymer physics. Particularly valuable is the treatment of path integrals in curved spaces and with topological constraints. The discussion of semiclassical approximations and their connection to classical mechanics illuminates the correspondence principle. Numerous worked examples guide readers through increasingly sophisticated calculations.

Kleinert, Hagen (2009). *Path Integrals in Quantum Mechanics, Statistics, Polymer Physics, and Financial Markets*, 5th Edition. Singapore: World Scientific.

This encyclopedic work demonstrates the universality of path integral methods across physics and beyond. Kleinert provides detailed derivations often sketched elsewhere and develops powerful approximation schemes. The book's treatment of path integrals with constraints, in spaces with curvature and torsion, and for systems with memory effects goes well beyond standard presentations. Applications to critical phenomena, superfluid helium, and even financial markets illustrate the method's breadth. The extensive bibliography and historical notes make this an invaluable reference for researchers.

Field Theory and Gauge Theories

Weinberg, Steven (1995-2000). *The Quantum Theory of Fields*, Volumes I-III. Cambridge: Cambridge University Press.

Weinberg's magisterial three-volume series provides the definitive modern treatment of quantum field theory from first principles. Volume I develops the general framework, emphasizing how quantum field theory emerges inevitably from combining special relativity with quantum mechanics. Volume II covers modern applications including the Standard Model, while Volume III treats supersymmetry. Throughout, Weinberg shows how the principle of least action, combined with symmetry principles, determines the structure of physical theories. His careful attention to foundations and clear physical reasoning make these challenging volumes accessible to dedicated students.

Peskin, Michael E. and Schroeder, Daniel V. (1995). *An Introduction to Quantum Field Theory*. Reading, MA: Addison-Wesley.

This pedagogical masterpiece has become the standard first-year graduate text in quantum field theory. Starting from the principle of least action for fields, Peskin and Schroeder develop quantum field theory with exceptional clarity. The book excels in its step-by-step calculations and physical interpretations of mathematical formalism. Detailed treatments of renormalization, gauge theories, and spontaneous symmetry breaking prepare students for research. The problems, ranging from straightforward to research-level, reinforce concepts and develop calculational skills.

Zee, A. (2010). *Quantum Field Theory in a Nutshell*, 2nd Edition. Princeton: Princeton University Press.

Zee's unique text conveys the conceptual beauty of quantum field theory while maintaining mathematical integrity. Using the path integral approach throughout, Zee shows how diverse phenomena—from Anderson localization to the cosmological constant problem—connect through field theoretic thinking. His conversational style and emphasis on physical intuition make difficult concepts accessible. The book's treatment of effective field theories and the renormalization group provides modern perspectives often missing from formal presentations. Extensive applications to condensed matter physics demonstrate field theory's relevance beyond particle physics.

Mathematical Methods

Gelfand, I.M. and Fomin, S.V. (1963). *Calculus of Variations.* Englewood Cliffs: Prentice-Hall. (Reprinted by Dover, 2000)

This classic text remains the standard mathematical reference for variational calculus. Gelfand and Fomin develop the subject rigorously while maintaining clear connections to physical applications. Their treatment of sufficient conditions for extrema, fields of extremals, and direct methods provides tools essential for modern applications. The book's careful attention to functional analysis foundations prepares readers for infinite-dimensional generalizations. Problems ranging from classical mechanics to optimal control theory illustrate the methods' versatility.

Courant, R. and Hilbert, D. (1953-1962). *Methods of Mathematical Physics*, Volumes I-II. New York: Interscience.

This monumental work shaped mathematical physics education for generations. Volume I covers variational methods extensively, including eigenvalue problems and direct methods that presage modern computational approaches. Volume II's treatment of partial differential equations includes extensive applications of variational principles. The books' influence extends beyond their specific content—their integration of rigorous mathematics with physical insight established a model for mathematical physics texts. Modern readers still find fresh insights in these carefully crafted volumes.

Olver, Peter J. (1993). *Applications of Lie Groups to Differential Equations*, 2nd Edition. New York: Springer-Verlag.

Olver's text provides the definitive treatment of symmetry methods in differential equations, with extensive applications to variational problems. The systematic development of prolongation theory, invariant solutions, and conservation laws through Noether's theorem makes this essential reading for understanding symmetries in physics. The book's examples span fluid mechanics, elasticity, quantum mechanics, and relativity. Clear exposition and numerous exercises guide readers from basic concepts to research-level techniques.

Specialized Topics

MacKay, R.S. and Meiss, J.D. (eds.) (1987). *Hamiltonian Dynamical Systems: A Reprint Selection.* Bristol: Adam Hilger.

This carefully curated collection of seminal papers traces the development of modern Hamiltonian dynamics. Including works by Poincaré, Birkhoff, Kolmogorov, Arnold, and Moser, the volume provides primary sources for KAM theory and chaos in Hamiltonian systems. Editorial introductions place papers in historical and scientific context, making classical results accessible to modern readers. The collection's emphasis on the interplay between regular and chaotic motion illuminates a central theme in nonlinear dynamics.

Tabor, Michael (1989). *Chaos and Integrability in Nonlinear Dynamics.* New York: Wiley-Interscience.

Tabor's text bridges classical integrability theory and modern chaos studies, showing how both emerge from Hamiltonian systems. The book excels in its treatment of soliton theory, inverse scattering methods, and connections to algebraic geometry. Clear explanations of technical topics like Bäcklund transformations and Painlevé analysis make advanced methods accessible. The integration of analytical and numerical approaches reflects contemporary research practice in nonlinear dynamics.

Rañada, Alberto F. (2014). *Dynamical Systems, Analytical Mechanics and Geometric Phases.* New York: Nova Science Publishers.

This modern monograph explores geometric aspects of classical mechanics often overlooked in standard texts. Rañada's treatment of geometric phases, including the Hannay angle and its quantum counterpart the Berry phase, reveals deep connections between classical and quantum mechanics. The book's discussion of reduction theory and momentum maps provides tools for analyzing systems with symmetry. Applications to molecular dynamics and celestial mechanics demonstrate the practical importance of geometric methods.

Modern Research and Reviews

Brizard, Alain J. (2023). *An Introduction to Lagrangian Mechanics*, 3rd Edition. Singapore: World Scientific.

This recent text provides a fresh perspective on Lagrangian mechanics, emphasizing geometric structures and modern applications. Brizard's treatment of nonholonomic systems, vakonomic mechanics, and field theories reflects current research interests. The book's careful attention to variational principles with constraints addresses subtleties often glossed over in standard presentations. Extensive examples from plasma physics and accelerator dynamics demonstrate the formalism's contemporary relevance.

Hamill, Patrick (2014). *A Student's Guide to Lagrangians and Hamiltonians.* Cambridge: Cambridge University Press.

Written specifically for undergraduates, Hamill's guide provides an accessible introduction to analytical mechanics. Clear explanations, worked examples, and conceptual discussions help students grasp abstract concepts. The book's treatment of phase space, canonical transformations, and connections to quantum mechanics prepares students for advanced study. Careful attention to common misconceptions and mathematical prerequisites makes this an ideal supplementary text.

Sussman, Gerald Jay and Wisdom, Jack (2001). *Structure and Interpretation of Classical Mechanics.* Cambridge: MIT Press. (2nd Edition, 2015)

This innovative text uses functional programming to explore classical mechanics, providing new insights through computational approaches. The authors develop a precise language for describing mechanical systems, implement it in Scheme, and use it to investigate complex dynamics. Their treatment of perturbation theory, invariant curves, and chaotic dynamics benefits from exact arithmetic and symbolic computation. The book's philosophy—that programming and mathematics are complementary tools for understanding—offers a paradigm for 21st-century theoretical physics.

Historical and Philosophical Perspectives

Yourgrau, Wolfgang and Mandelstam, Stanley (1968). *Variational Principles in Dynamics and Quantum Theory*, 3rd Edition. London: Pitman. (Reprinted by Dover, 1979)

This unique work combines historical development with technical exposition, tracing variational principles from Fermat through quantum field theory. The authors' philosophical discussions of teleology, causality, and the nature of physical laws provide context often missing from technical texts. Their treatment of the action principle in general relativity and early quantum field theory captures the state of knowledge before modern developments. The book remains valuable for its integrated approach to history, philosophy, and physics.

Coopersmith, Jennifer (2017). *The Lazy Universe: An Introduction to the Principle of Least Action.* Oxford: Oxford University Press.

Coopersmith's recent book provides an accessible yet thorough introduction to the principle of least action for non-specialists. Combining historical narrative with clear explanations, she traces the principle's development from ancient philosophy through modern physics. The book's strength lies in its conceptual discussions and careful attention to common misunderstandings. Extensive references and suggestions for further reading make this an excellent starting point for students and educators.

Stöltzner, Michael (1994). "Action Principles and Teleology." In *Inside Versus Outside*, edited by H. Atmanspacher and G.J. Dalenoort, pp. 33-62. Berlin: Springer.

This scholarly analysis examines the philosophical implications of action principles, addressing longstanding concerns about their apparently teleological nature. Stöltzner traces debates from Maupertuis through modern interpretations, showing how mathematical equivalence does not resolve philosophical questions. His discussion of competing interpretations—causal, teleological, and structural—illuminates ongoing foundational debates. The paper's extensive bibliography provides entry to the philosophical literature on variational principles.

Computational Methods

Leimkuhler, Benedict and Reich, Sebastian (2004). *Simulating Hamiltonian Dynamics.* Cambridge: Cambridge University Press.

This monograph addresses the crucial problem of numerically integrating Hamiltonian systems while preserving their essential properties. The authors develop geometric integrators that maintain symplectic structure, ensuring long-term stability and conservation laws. Their treatment of constrained systems, multiple time scales, and stochastic perturbations addresses practical challenges in molecular dynamics and celestial mechanics. Clear exposition of backward error analysis and modified equations helps readers understand why structure preservation matters.

Hairer, Ernst, Lubich, Christian, and Wanner, Gerhard (2006). *Geometric Numerical Integration*, 2nd Edition. Berlin: Springer.

This comprehensive treatment of structure-preserving algorithms has become the standard reference for geometric integration. Starting from variational principles, the authors show how discrete mechanics leads naturally to symplectic integrators. Their analysis of energy conservation, momentum maps, and reversibility provides deep insight into algorithm behavior. Applications ranging from molecular dynamics to electromagnetic field computation demonstrate the methods' broad applicability. The book's rigorous mathematical approach rewards careful study.

Contemporary Frontiers

Rovelli, Carlo (2004). *Quantum Gravity.* Cambridge: Cambridge University Press.

Rovelli's text provides a comprehensive introduction to loop quantum gravity, one approach to quantizing general relativity. Starting from the Einstein-Hilbert action, he shows how diffeomorphism invariance leads to a radically different quantization than perturbative approaches. The book's careful development of spin networks, area operators, and quantum geometry makes cutting-edge research accessible. While advocating for loop quantum gravity, Rovelli fairly presents alternative approaches and open problems.

Polchinski, Joseph (1998). *String Theory,* Volumes I-II. Cambridge: Cambridge University Press.

Polchinski's texts have become the standard reference for string theory, providing rigorous treatment of a subject often presented heuristically. Starting from the Polyakov action, he develops the full machinery of perturbative string theory. Volume I covers bosonic strings, while Volume II treats superstrings and D-branes. The books' strength lies in their careful attention to technical details while maintaining physical insight. Extensive problems guide readers through calculations essential for research.

Becker, Katrin, Becker, Melanie, and Schwarz, John H. (2007). *String Theory and M-Theory: A Modern Introduction.* Cambridge: Cambridge University Press.

This modern text emphasizes string theory's connections to observable physics, including applications to cosmology and particle phenomenology. The authors develop the subject from the Nambu-Goto and Polyakov actions, emphasizing how spacetime physics emerges from worldsheet dynamics. Their treatment of flux compactifications, moduli stabilization, and the landscape addresses contemporary research directions. Clear physical discussions balance technical developments, making this accessible to students with quantum field theory background.

This bibliography represents a selection of essential works on the principle of least action and its ramifications throughout physics. From historical foundations through contemporary frontiers, these references provide pathways for deeper exploration of one of physics' most profound organizing principles. The literature continues to grow as researchers discover new applications and interpretations of variational methods, ensuring that the principle of least action remains at the heart of theoretical physics.

Index

C

D

E

F

G

H

I

N

O

P

Q

R

S

T

U

V

W

Y

Z

www.ingramcontent.com/pod-product-compliance
Lightning Source LLC
Chambersburg PA
CBHW041747220926
51847CB00013B/134
* 9 7 8 2 4 8 8 6 1 6 0 0 3 *